The QS 9000 Documentation Toolkit

The ISO Solutions Series
by Janet L. Novack

The ISO 9000 Documentation Toolkit

The ISO 9000 Quality Manual Developer

The QS 9000 Documention Toolkit
(Janet L. Novack and Kathleen C. Bosheers)

The QS 9000 Documentation Toolkit

Janet L. Novack
and
Kathleen C. Bosheers

To join a Prentice Hall PTR internet mailing list, point to http://www.prenhall.com/register

Prentice Hall PTR
Upper Saddle River, New Jersey 07458
http://www.prenhall.com

Library of Congress Cataloging-in-Publication Data

Novack, Janet L.
The QS 9000 documentation toolkit / Janet L. Novack and Kathleen C. Bosheers.
p. cm.
Includes index.
ISBN 0-13-653643-3 (alk. paper)
1. QS 9000 (Standard) 2. Automobile industry and trade—Quality control—Standards—United States.
I. Bosheers, Kathleen C. II. Title
TL278.N68 1997
629.23'4'021873--dc20 96-35961
CIP

Editorial/Production Supervision: Kathleen M. Caren
Acquisitions Editor: Bernard M. Goodwin
Janet Novack's photographer: Robert W. Service
Manufacturing Manager: Alexis R. Heydt
Cover Design Director: Jerry Votta

Prentice-Hall, Inc.
A Simon & Schuster Company
Upper Saddle River, NJ 07458

The publisher offers discounts on this book when ordered in bulk quantities.
For more information, contact:

Corporate Sales Department
Prentice Hall PTR
1 Lake Street
Upper Saddle River, NJ 07458
Phone: 1-800-382-3419
FAX: 201-236-7141
email: corpsales@prenhall.com

Printed in the United States of America

10 9 8 7 6 5 4 3 2 1

ISBN 0-13-653643-3

Prentice-Hall International (UK) Limited, *London*
Prentice-Hall of Australia Pty. Limited, *Sydney*
Prentice-Hall of Canada, Inc., *Toronto*
Prentice-Hall Hispanoamericana S.A., *Mexico*
Prentice-Hall of India Private Limited, *New Delhi*
Prentice-Hall of Japan, Inc., *Tokyo*
Simon & Schuster Asia Pte. Ltd., *Singapore*
Editora Prentice-Hall do Brasil, Ltda., *Rio de Janeiro*

TABLE OF CONTENTS

Preface

In August 1994, the Automotive Industry Action Group published *Quality System Requirements, QS 9000.* Representatives from the automotive industry had worked over four years to develop a quality system requirement that used the *ISO 9001, 1994 Edition* as a baseline.

The *QS 9000* supplements the ISO 9001 with automotive-specific requirements. Those requirements are intended to enhance overall automotive quality while reducing variation and waste throughout the entire supply chain. In February 1995, a second edition of the *QS 9000* was published to clarify the intent of the manual and to keep the manual current.

The **QS 9000 Documentation Toolkit** reflects the February 1995 edition of the *QS 9000.* We also used the supporting documents of the *QS 9000* in developing the worksheets and procedures. These documents are:

- *Advanced Product Quality Planning and Control Plan Reference Manual*
- *Production Part Approval Process*
- *Potential Failure Mode and Effects Analysis*
- *Measurement Systems Analysis*
- *Statistical Process Control Reference Manual*

These documents are available from the Automotive Industry Action Group. You can order the documents by calling (810) 358-3003. We suggest that you obtain a full set of these documents and read them before starting your QS 9000 efforts.

The *QS 9000* is a quality system requirement for internal and external suppliers of production parts, service parts, and finishing services (heat treat, paint, etc.). In other words, if you supply product directly to General Motors, Chrysler, or Ford that is put on a car, you need to comply with the QS 9000. If you are a subcontractor to a supplier that furnishes product directly to General Motors, Chrysler, or Ford, your customer may require you to comply to the QS 9000. In either case, if you have any questions regarding the applicability of your product, contact your customer's quality representative or purchasing office.

Just like the ISO 9000 Documentation Toolkit, this book is designed to evoke discussion at planning meetings, to be annotated and written in, and to be employed in the writing of procedures.

Basically, this book has two parts-the worksheets and the software procedures. We are providing the software procedures in WordPerfect® for Windows 6.0, Microsoft Word® for Windows 6.0, and Microsoft Word® for Macintosh 6.0. Hopefully, you can use one of these formats. If not, this book contains the printed version of the procedures.

Just one note, the diskette is formatted for MS-DOS. So, if you are using it with a Macintosh computer, you will need software installed on your Macintosh that reads MS-DOS disks.

When we decided to provide the **QS 9000 Documentation Toolkit**, we wanted all of you who had previously used the ISO 9000 Documentation Toolkit, to be able to quickly identify the areas that require change to be QS 9000-compliant. Therefore, we identified QS 9000-specific areas in normal type and ISO 9000 areas in italic type-this is consistent with the format of the *QS 9000.* If you already have an ISO 9000-based system, you can focus on the QS 9000-specific requirements. If you are pursing QS 9000 without prior ISO experience, you can use all the information and readily differentiate the two documents.

A word of advice for all of you pursuing QS 9000. Understand your processes and your product development system. If you do not understand your own product development system, developing an implementation strategy for QS 9000 will be difficult and overly time-consuming. The Advanced Product Quality Planning and Control Plan Reference Manual defines the customer's requirements for quality planning, but you must incorporate these and the customer's program timing requirements into your product development cycle. Remember, a basic tenet of any successful quality system is to understand your processes.

The approach used in this book demands that a company work as a team in defining its quality system. The worksheets interpret the *QS 9000* by specifying what you need to do and also demonstrate real-life strategies. They force you to think about who's responsible for an activity, the recordkeeping, and follow-up activities. The worksheets were originally developed to gather data. As they evolved, the worksheets became a tool that facilitated discussion and encouraged employee participation. We've been involved in lively discussions when staff at all levels contributed to the worksheets. Companies have discovered that open communication and the opportunity to improve how things were done became one of the unmeasured benefits of seeking QS 9000 certification.

Putting together a documented system not only requires time to understand the standard and a commitment to implement changes, but the skills to write it all down.

Word-processor templates in both Microsoft Word for IBM-compatibles or Macintoshes and in WordPerfect for Windows have been derived from the worksheets. You can simply edit the formatted templates to create your QS 9000 documentation. The templates include the elements of document control required by the ISO 9001 and QS 9000. Using the master template, you can write individual procedures or split the ones provided with this book. The templates contain more detail than you'll need on the assumption that it is easier to use the DELETE key than to create original text. You may find, however, that you need to expand on the procedures because of the complexity of your organization or product. If you do, remember, procedures need to identify who, what, when, and where.

This book provides the tools to produce your ISO 9000 and QS 9000 Level B quality procedures. To comply with the standard, you will also need to write a Quality Manual, quality plans including control plans, and the appropriate work instructions.

We need to thank the many people and organizations that have helped us assure the usefulness of the **QS 9000 Documentation Toolkit**. Our thanks to Larry Rainwater, Jan Black, Brenda Clifton, and Don Turner for helping us validate the QS 9000-specific questions in the worksheets; Nikki Hornbacker and the staff of Progressive Strategies & Systems Inc. for their support; and, Sue Miller for helping us format this book. Our families have unselfishly given us the encouragement and time to write and edit this book. (Thanks Pat, Grady, and Jodie!)

Finally, we hope this book will help you develop a system that conforms to your customer's requirements and helps you implement a functional quality system to support your efforts well into the next century.

Kathleen C. Bosheers
Phone: 810-781-2235
Fax: 810-781-4596
email: kbosheers@aol.com

From Janet Novack

When Kathy Bosheers of Progressive Strategies & Systems, Inc. expressed an interest in expanding *the* ISO 9000 Documentation Toolkit to include the QS 9000, she probably did not know what she was getting into. What she did know was that her clients found the ISO 9000 Documentation Toolkit a benefit to jump-starting implementation of their quality system. These clients, however, needed more. They were going for QS 9000. The enormous amount of time and effort Kathy and her staff spent to make the **QS 9000 Documentation Toolkit** a reality is evident in the completeness of this text. I am proud to add this book to the ISO solutions series. Were it not for Kathy, this book would not have been written.

Future plans for a QS 9000 Quality Manual Developer to help write the quality manual for QS 9000 worksheets, which can be filled in with your word processor and document control software are in the works. To find out more, complete the registration card at the back of this book and send it to Janet Novack, 38 Hamlet Street, Newton, MA, 02159.

Janet L. Novack
Phone: 617-965-5115
Fax: 617-969-7273
email: novackj@aol.com

Introduction

1

What's in This Section?

This section:

- explains the concept of this book and how to use it
- explains where to order the QS 9000 and supporting documents
- explains the worksheets and procedure templates for your word processor

About This Book

This book takes you through the process of defining your QS 9000 implementation plans and writing the required documentation. Table 1-1 describes the organization of this book.

Section	Description
Section 1, Introduction	Describes the worksheets and procedure templates, and portrays a finished QS 9000 procedure. This section also illustrates the worksheets and templates and highlights their features.
Section 2, Planning Your QS 9000 System and Documentation	Defines the types of QS 9000 documentation and provides a strategy for developing the documentation. This section helps you define the procedures that document your QS quality system.
Section 3, Writing the Procedures	Explains how to use the word processor text files to create the QS 9000 documentation. This section includes tips and techniques specific to using the text files.
Section 4, Worksheets	Holds the 24 worksheets for you to complete and 23 procedure templates. One worksheet, "Customer Specific Requirements," helps you identify your customer's requirements in relation to the other elements of the QS 9000.
Appendix A	Explains how to install the software, describes the directory and file structure, and teaches you how to open files. This section has instructions for disks formatted for Microsoft Word for Windows®, WordPerfect®, and Microsoft Word for Macintosh®.
Appendix B	Provides a form for you to use to track your progress.
Appendix C	Lists the QS 9000 quality system accreditation bodies.
Index	
Order Form	Provides an order form to request more information and to order Macintosh-formatted diskettes.
Templates	Contains the computer disks with files that you edit to document your QS 9000 system.

Table 1-1. Section Descriptions

QS 9000 Documents

The QS 9000 must be obtained from the Automotive Industry Action Group (AIAG). You can order the QS 9000 and supporting documents by calling the AIAG's Order Desk at (810) 358-3003.

The supporting documents to the QS 9000 are:

- Advanced Product Quality Planning (APQP) and Control Plans
- Potential Failure Mode and Effects Analysis
- Production Part Approval Process
- Measurement Systems Analysis
- Statistical Process Control Reference Manual

You will need to develop procedures for Advanced Product Quality Planning and Production Part Approval. We have included a worksheet and template for Production Part Approval.

An effective approach to Advanced Product Quality Planning should encompass a product development system. Read the APQP and adapt the requirements to your product development system. Then, document the appropriate procedures.

The Potential Failure Mode and Effects Analysis, Measurement Systems Analysis, and Statistical Process Control manuals provide guidance for using these tools as part of your quality and product development system.

The International Automotive Sector Group (IASG) is an international working group that resolves interpretation issues regarding the QS 9000. The results of the IASG are published by the American Society of Quality Control (ASQC). These interpretations provide you with clarification about the QS 9000 and are considered binding by Chrysler, General Motors, and Ford. The worksheets in this book were developed using the interpretations dated March 22, 1996. We suggest you obtain a copy of the IASG interpretations and review them when you are developing your quality system. The interpretations can be ordered by calling the ASQC at 1-800-248-1946. You can also access the interpretations from the ASQC QS 9000 Web Site at http:/asqc.org/9000.

Overview

The **QS 9000 Documentation Toolkit** will aid your understanding of the activities required for certification by identifying specific requirements. While we've covered many circumstances in the **QS 9000 Documentation Toolkit**, you may, however, have questions that apply to your particular situation. We encourage you to seek assistance from a QS 9000 consultant or from the registering body to whom you are applying for certification.

The QS 9000 standard is broadly written to address the basic requirements of a quality system that provides parts to General Motors, Chrysler, or Ford. The standard does not tell you HOW to meet the requirements; it only describes the requirements. How you meet a requirement depends on what works best in your environment. If you currently have a quality system meeting an ISO requirement, you may find the QS 9000 to be more prescriptive.

This book outlines suggested activities to comply with the standard. The activities are guidelines to help you understand and follow an implementation strategy.

The **QS 9000 Documentation Toolkit** helps you understand how to apply the QS 9000 standard to your daily tasks and saves you time in documenting your quality system by:

- providing detailed worksheets
- clarifying the required activities with real-life examples
- providing word processor templates to convert your step-by-step plans into the required documentation

The Planning Process

During the planning stage, it's important that all members of the QS 9000 team know and understand the scope of the implementation plan.

The **QS 9000 Documentation Toolkit** provides step-by-step guidance for implementing the QS 9000 and helps ensure the activities receive the required interdepartmental follow-up. Build your implementation plans from the worksheets. The worksheets help you identify who's responsible for an activity, what records are kept, and how the activity is reviewed. After consensus on the worksheets, edit the procedure templates on your word processor to create the required documentation. Figure 1-1 illustrates the planning process flow.

Figure 1-1. Process of Creating QS 9000 Procedures

Understanding the Worksheets

The worksheets facilitate the data-gathering process for defining your QS 9000 plans. There are 24 worksheets corresponding to the 24 QS 9000 elements. The "Customer Specific Requirements" worksheet identifies the Chrysler, General Motors, and Ford requirements to the other QS 9000 elements. The intent of this worksheet is for you to incorporate your customer's requirements into the appropriate QS 9000 element procedure.

The QS 9000 is written in general terms, but you must adhere to specific requirements. The QS 9000 tells you WHAT you need to do to comply. It defines supporting strategies but it does not tell you exactly HOW to do it. The HOW is for you to define and document.

The intent of the worksheets is to demystify the QS 9000 and provide you with a proven method of developing your plans.

1. Start by reviewing your current activities.

2. Then, determine where the activity fits into the standard.

3. Go through the worksheets to gain an understanding of which activities you perform and where in the standard they are addressed.

4. Determine which QS requirements you do not meet. Where you are not currently in compliance with the standard, the worksheets will demonstrate examples of activities that have been employed by registered companies.

What's in a Worksheet?

The worksheets are loaded with questions to help you elicit responses. The format of the questions are designed to be easy to answer and use:

- yes/no questions to highlight the required activities
- fill-in-the-blanks to establish the specifics of your process
- multiple choice questions to offer implementation suggestions

The worksheets provide the necessary interdepartmental communication by helping you identify follow-up activities and ensure that all gaps in your quality system are closed. When completing the worksheets, consider who's responsible for a task and what records are generated.

Common to each worksheet are learning aids designed to promote your understanding of the QS 9000. Table 1-2 explains these aids. Figures 1-2 and 1-3 illustrate sample pages from the worksheets and identify the learning aids. The QS 9000 document differentiates ISO 9001 from QS 9000-specific requirements by using different type faces. **The worksheets and procedures in the book use this same methodology: QS 9000 is in normal type and ISO 9001 information is in italic type.**

Following each worksheet is a printed version of the procedure template that you will find on the disk included at the back of this book. The printed version does not show the procedure's cover page, header, and footer. These elements, however, are part of the software version. Additionally, all procedures in the disk version are in normal type. Figure 1-4 illustrates an edited printout of the software version, including the cover page, header, and footer.

Aid	What it Means
ISO Standard	Refer to the standard for clarification when developing your implementation plans.
QS 9000 Interpretations and Supplemental Quality System Requirements	This section summarizes the QS 9000 additions to the ISO 9001 element. You will need to read the actual QS 9000 requirements when completing the worksheets.
Suggested procedures	You may want to segment the worksheets into several procedures. This aid suggests possibilities. Add procedures as appropriate for your company.
Examples for implementation	As you read the questions, you will find strategies that have been successfully implemented by ISO 9000-certified companies and companies pursing QS 9000 certification.
Margin notes	The worksheets detail activities beyond the requirements of QS 9000. The margin notes specify what is required by the ISO 9001 and QS 9000. Use the margin notes to distinguish a QS 9000 requirement from a suggested implementation strategy. Where you see a note, you must define how you meet the requirement.
Implementation aids	Within the text of the worksheets you will be asked to assign a person responsible for a task and to identify any records generated by the task.

Table 1-2. Worksheet Learning Aids

Figure 1-2 illustrates the first page of the Contract Review worksheet. Notice that the relevant standard is printed on this page. Use it as a handy reference when completing worksheets. Also on the page of the worksheet is a suggested procedure list. Use this as a guide when identifying the procedures you will be documenting.

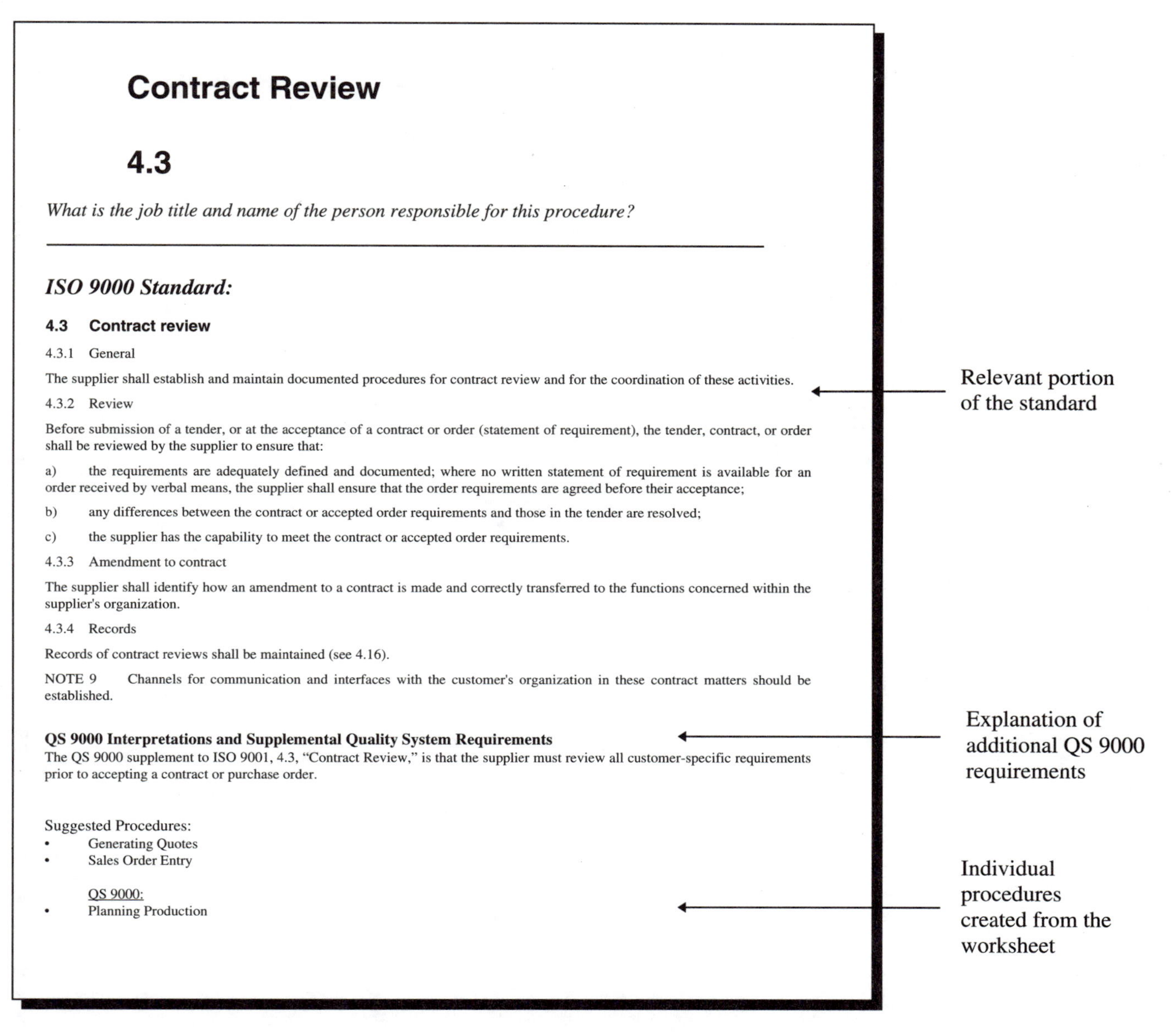

Contract Review

4.3

What is the job title and name of the person responsible for this procedure?

ISO 9000 Standard:

4.3 Contract review

4.3.1 General

The supplier shall establish and maintain documented procedures for contract review and for the coordination of these activities.

4.3.2 Review

Before submission of a tender, or at the acceptance of a contract or order (statement of requirement), the tender, contract, or order shall be reviewed by the supplier to ensure that:

a) the requirements are adequately defined and documented; where no written statement of requirement is available for an order received by verbal means, the supplier shall ensure that the order requirements are agreed before their acceptance;

b) any differences between the contract or accepted order requirements and those in the tender are resolved;

c) the supplier has the capability to meet the contract or accepted order requirements.

4.3.3 Amendment to contract

The supplier shall identify how an amendment to a contract is made and correctly transferred to the functions concerned within the supplier's organization.

4.3.4 Records

Records of contract reviews shall be maintained (see 4.16).

NOTE 9 Channels for communication and interfaces with the customer's organization in these contract matters should be established.

QS 9000 Interpretations and Supplemental Quality System Requirements
The QS 9000 supplement to ISO 9001, 4.3, "Contract Review," is that the supplier must review all customer-specific requirements prior to accepting a contract or purchase order.

Suggested Procedures:

- Generating Quotes
- Sales Order Entry

QS 9000:

- Planning Production

Figure 1-2. First Page of Worksheet

Figure 1-3 illustrates a sample page from the Management Responsibility worksheet. The text in the margin identifies what's required by the standard. Use these notes to distinguish between what you must implement for QS 9000 and what would be nice to have.

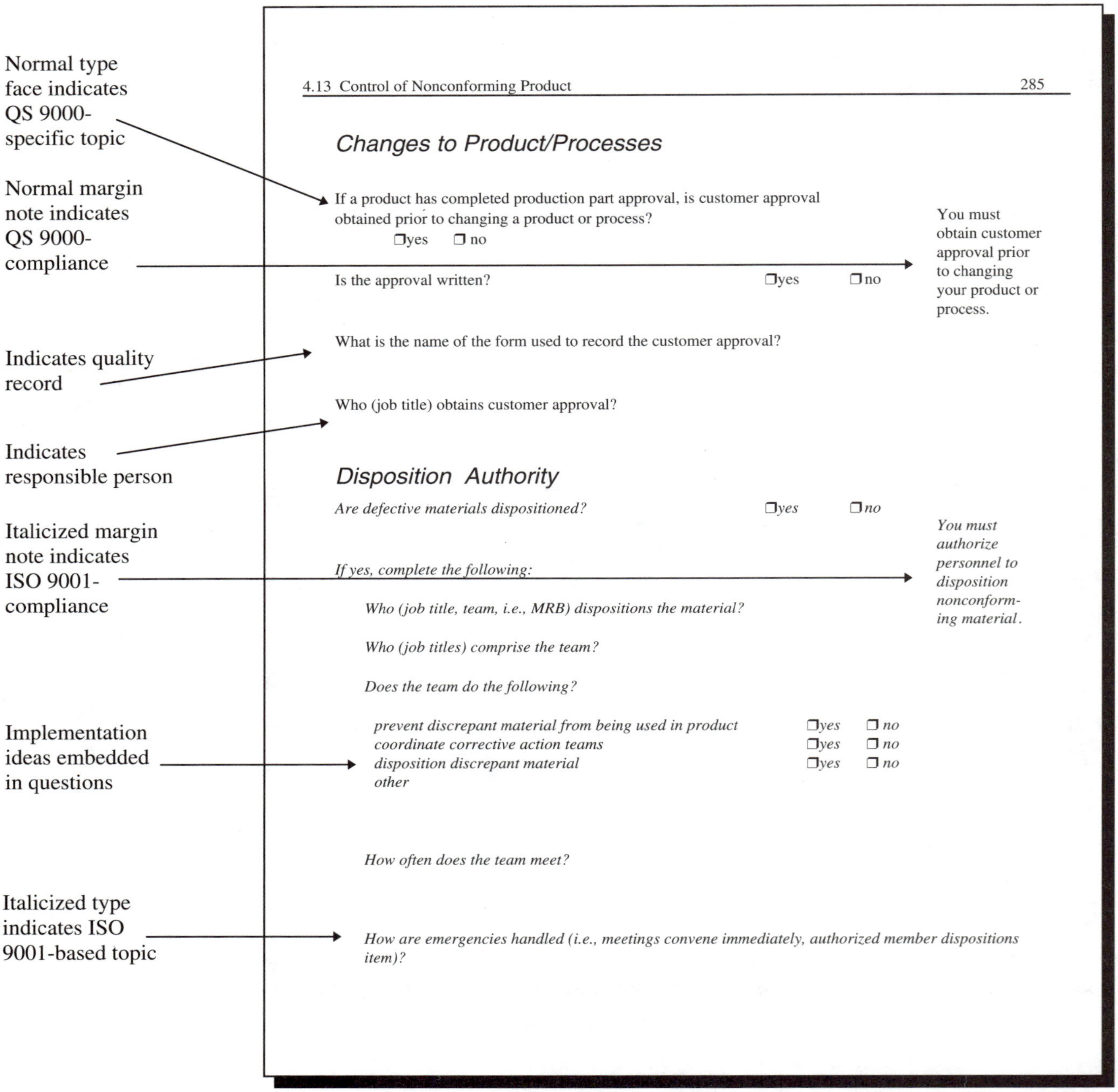

Normal type face indicates QS 9000-specific topic

Normal margin note indicates QS 9000-compliance

Indicates quality record

Indicates responsible person

Italicized margin note indicates ISO 9001-compliance

Implementation ideas embedded in questions

Italicized type indicates ISO 9001-based topic

4.13 Control of Nonconforming Product 285

Changes to Product/Processes

If a product has completed production part approval, is customer approval obtained prior to changing a product or process?
❐yes ❐ no

You must obtain customer approval prior to changing your product or process.

Is the approval written? ❐yes ❐no

What is the name of the form used to record the customer approval?

Who (job title) obtains customer approval?

Disposition Authority

Are defective materials dispositioned? ❐*yes* ❐*no*

You must authorize personnel to disposition nonconforming material.

If yes, complete the following:

Who (job title, team, i.e., MRB) dispositions the material?

Who (job titles) comprise the team?

Does the team do the following?

prevent discrepant material from being used in product ❐*yes* ❐ *no*
coordinate corrective action teams ❐*yes* ❐ *no*
disposition discrepant material ❐*yes* ❐ *no*
other

How often does the team meet?

How are emergencies handled (i.e., meetings convene immediately, authorized member dispositions item)?

Figure 1-3. Sample Worksheet Page

Understanding the Procedure Templates

The procedure templates are compatible with Macintosh and personal computers and are available in Microsoft Word and WordPerfect formats.

Appendix A provides information about installing the software, understanding the directory and file structures, and opening files.

Procedure templates include:

- 23 procedure templates, one for each element of the QS 9000, except "Customer-Specific Requirements."
- One master template to customize the scope and content of your documentation.

We supplied the templates to help you write your drafts. The templates are abundant with information, with the belief that it is easier to DELETE than to create new text.

Table 1-3 identifies what's common to the procedure templates. This includes elements of document control, section titles, and textual information. Figures 1-4 through 1-6 illustrate pages from the templates and indicate the elements described in Table 1-3.

Topic	Element	What it Means
Document Control	Document number	Identifies the procedure with a unique number.
	Revision number	Identifies the revision level.
	Date	Identifies the release date of the procedure. This is automatically updated each time you print the procedure.
	Page count	Shows the page number and total page count in the form: X of Y. Page numbering is automatic; however, you must enter the last page in place of 'Y.'
	Approval signatures	Provides an area for approval signatures. Revisions must be approved by the original approver or designee.

Table 1-3. Elements of Procedure Templates

Topic	Element	What it Means
	Change record	Describes modifications to a revised procedure. This may appear on the document itself, or you may choose to keep this record as a separate document.
	Distribution list	Lists holders of controlled copies. This may appear on the document itself, or you may choose to keep this record as a separate document. Use this list when replacing obsolete copies with the revised version.
	Controlled copy identification	Identifies the document as a controlled copy. This must appear on all controlled copies and must be distinguishable from photocopies. We suggest the use of a color code on controlled copies that cannot be mechanically duplicated.
Section titles	Purpose	States the purpose of the procedure.
	Scope	States the scope of the procedure and may include business lines, products, facilities, and exclusions, if any.
	Responsibilities	Identifies, by job title, employees who perform activities described in the procedure.
	Procedure	Describes the activities and tasks performed to meet the QS 9000 requirement.
	Related documents	Lists quality records, work instructions, and other procedures mentioned or associated with this procedure.
Textual information	Text	Derived from the worksheets, the text indicates placement for worksheet responses.
	Variable text, in angled brackets	Substitute worksheet responses between angled brackets.
	Cross references	Identify related procedures that provide more information.

Table 1-3. Elements of Procedure Templates (Continued)

Figure 1-4 shows the first page of a procedure template. Notice the elements of document control.

Company Name Division or Address Division or Address	**Management Responsibility**	
	Doc. No.	Rev. No.
	Date: 6/6/96	Page 1 of x

your logo

QS Procedure

Management Responsibility

Approved: ______________________
title Name

Approved: ______________________
title Name

Approved: ______________________
title Name

Approved: ______________________
title Name

Approved: ______________________
title Name

Change Record

Rev	Date	Responsible Person	Description of Change
		Name	Initial Release

Distribution List
(list the departments that receive controlled copies)

Controlled Copy, Do Not Duplicate **For Internal Use Only**

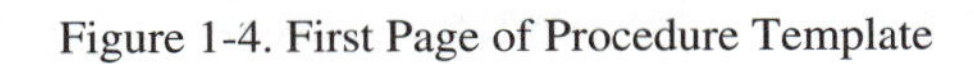

Figure 1-4. First Page of Procedure Template

Figure 1-5 shows the second page of a procedure template. It includes sections for purpose, scope, and responsibilities.

Purpose explains why. If you find you have more purpose statements than you need, delete unnecessary ones.

Scope identifies areas covered in the procedure

Responsibilities assigns action to tasks

Company Name	**Management Responsibility**	
Division or Address	Doc. No.	Rev. No.
Division or Address	Date:	Page 2 of x

1. Purpose

- To establish a mechanism for management to set quality goals and review their attainment.
- To provide for management review of internal quality audits and completion of corrective action.
- To ensure all employees are adequately trained.
- To establish and implement a quality system by formulating the quality policy, defining organization, assigning authorities and responsibilities, appointing the management representative, reviewing the quality system, and ensuring the resources and personnel to maintain the system.
- To clearly identify the management function responsible for the quality system.
- To establish a quality policy and objectives and ensure that the policy is communicated and practiced throughout the organization and that quality objectives are routinely met by everyone.
- To have in place an organization chart and a description of roles and responsibilities required for the business unit. The organization chart is a controlled document and updated at least once a year.
- To establish and implement a system for strategic planning through the use of a documented Business Plan. The Business Plan includes the analysis of company-level data, benchmarked information, and customer satisfaction indicators.

2. Scope

This procedure applies to all operations at <Company>.

3. Responsibilities

The <job title> has overall responsibility for quality at <Company>.

The QS 9000 Team, composed of the <job titles>, is responsible for forming an internal audit team to audit QS 9000-compliance.

The Senior Management Team, composed of <job titles>, analyzes company-level data and customer satisfaction indicators.

The <job title> prepares, updates, and revises the business plan.

The <job title> approves the Business Plan.

The <job title, team name> reviews the Business Plan.

Management supports the organizational freedom and authority to:

- Identify and record any product quality problems.
- Initiate, recommend, or provide solutions through designated channels.
- Verify the implementation of solutions.
- Control further processing, delivery, or installation of nonconforming product until the deficiency or unsatisfactory condition has been corrected.

Controlled Copy, Do Not Duplicate **For Internal Use Only**

Figure 1-5. Second Page of Procedure Template

Figure 1-6 is a sample page of a procedure. The procedure templates were derived directly from the worksheets. Substitute worksheet entries for generic information in angled brackets (< >).

Company Name	**Management Responsibility**	
Division or Address	Doc. No.	Rev. No.
Division or Address	Date:	Page 1 of x

4.11 Customer Satisfaction

See *Continuous Improvement, xxx*

See *Corrective and Preventive Action, xxx*

The <job title> is responsible for conducting <frequency> customer satisfaction assessments. <Job title> maintains the results of the customer assessments.

The results of the assessments are reviewed by the <Senior Management Team> to identify trends in customer satisfaction and indicators of customer dissatisfaction. Industry benchmarks are used when reviewing the customer assessments.

Pertinent information from the customer assessments are used to initiate corrective action. The information is used during the review, update, and revision of the <Company> Business Plan.

4.12 Business Plan

<Job title, Team Name> is responsible for the preparation and revision of the <Company> Business Plan. The Business Plan contains short-term and long-term goals for maintaining customer satisfaction.

The Business Plan is reviewed by <job title, Team Name> and updated by <job title>. Review and update of the Business Plan is done, at a minimum, <frequency>.

Customer expectations, trends, industry benchmarks, and world-class benchmarks are used to develop, update, and revise the Business Plan.

The Business Plan is maintained by <job title> and may be reviewed with customers by <job title> on an individual basis.

<Job title, Team Name> updates the organization on the Business Plan <frequency>. The update is accomplished by:

<activity>

<activity>

5. Related Documents

<record of management review>
<training records>
<Business Plan>
<customer assessment records>
<company-level data records>

Corrective and Preventive Action, xxx
Internal Quality Audits, xxx
Quality Records, xxx
Training, xxx
Continuous Improvement, xxx

<list of work instructions>

Controlled Copy, Do Not Duplicate **For Internal Use Only**

Cross-references other procedures

Follows contents of worksheets

Identifies quality records

Figure 1-6. Sample Page, Management Responsibility Procedure Template

An Example of a Final Procedure

Use the templates as best fits your organization. Finished procedures can vary from the template in style, context, and sequence. These are your procedures, you must abide by them, and they work best when written, created, and maintained by your staff.

Above all,

SAY WHAT YOU DO, DO WHAT YOU SAY.

Figures 1-7 through 1-11 show an actual seven-page procedure created using the worksheet-to-template-to-finished-procedure method. Company references have been removed.

Company Name ABC Company Division or Address 1234 Anywhere St. Division or Address Anyplace, USA	**Management Responsibility**	
	Doc. No.	Rev. No.
	Date:	Page 1 of 5

your logo

QS Procedure
Management Responsibility

Approved: title	Name
Approved: title	Name
Approved: title	Name
Approved: title	Name
Approved: title	Name

Change Record	RevDate	Responsible Person	Description of Change
		Name	Initial Release

Distribution List

(list the departments that receive controlled copies)

Controlled Copy, Do Not Duplicate **For Internal Use Only**

Figure 1-7. Finished Procedure, Page 1

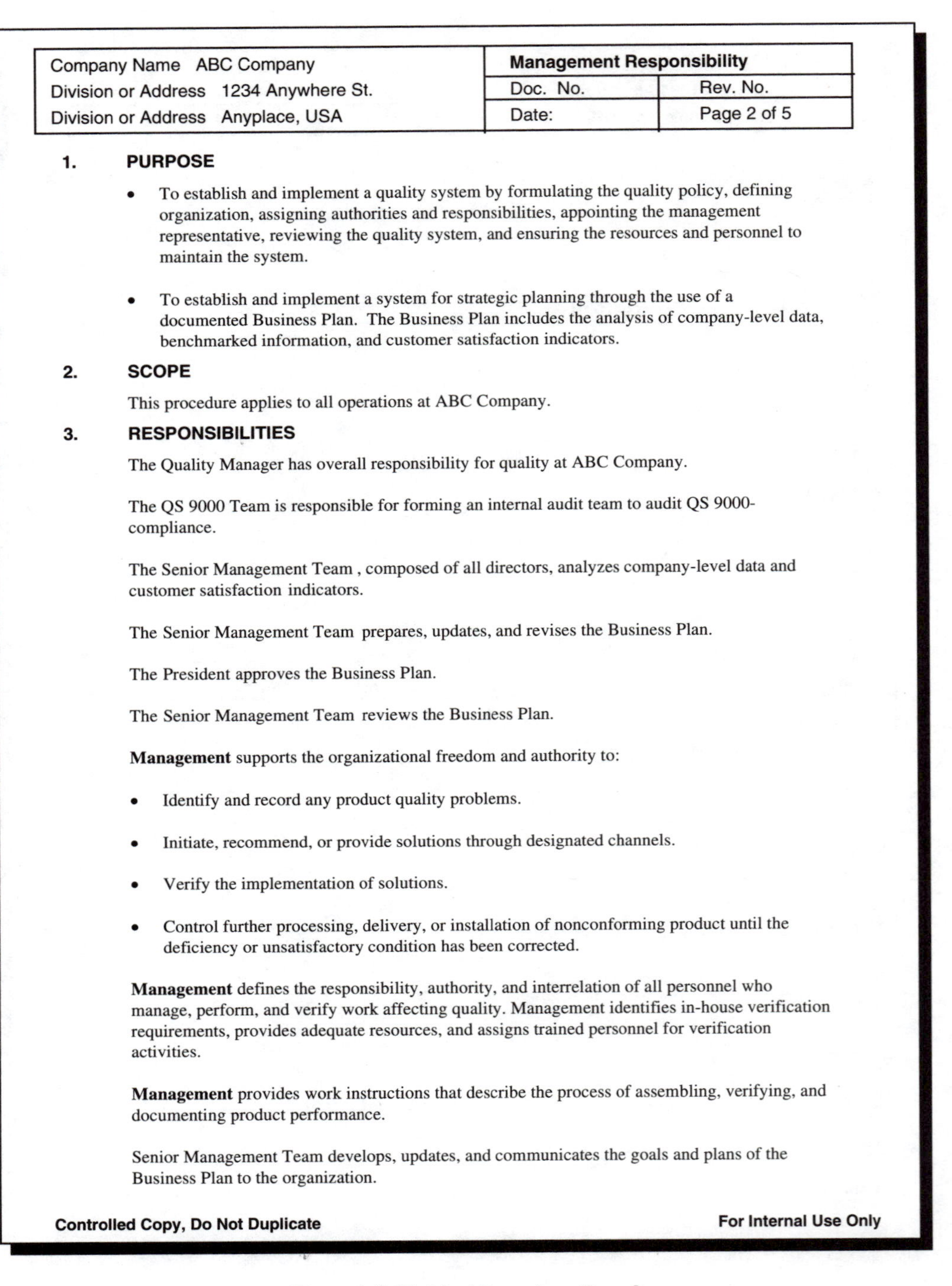

Company Name ABC Company Division or Address 1234 Anywhere St. Division or Address Anyplace, USA	**Management Responsibility**	
	Doc. No.	Rev. No.
	Date:	Page 2 of 5

1. PURPOSE

- To establish and implement a quality system by formulating the quality policy, defining organization, assigning authorities and responsibilities, appointing the management representative, reviewing the quality system, and ensuring the resources and personnel to maintain the system.

- To establish and implement a system for strategic planning through the use of a documented Business Plan. The Business Plan includes the analysis of company-level data, benchmarked information, and customer satisfaction indicators.

2. SCOPE

This procedure applies to all operations at ABC Company.

3. RESPONSIBILITIES

The Quality Manager has overall responsibility for quality at ABC Company.

The QS 9000 Team is responsible for forming an internal audit team to audit QS 9000-compliance.

The Senior Management Team , composed of all directors, analyzes company-level data and customer satisfaction indicators.

The Senior Management Team prepares, updates, and revises the Business Plan.

The President approves the Business Plan.

The Senior Management Team reviews the Business Plan.

Management supports the organizational freedom and authority to:

- Identify and record any product quality problems.
- Initiate, recommend, or provide solutions through designated channels.
- Verify the implementation of solutions.
- Control further processing, delivery, or installation of nonconforming product until the deficiency or unsatisfactory condition has been corrected.

Management defines the responsibility, authority, and interrelation of all personnel who manage, perform, and verify work affecting quality. Management identifies in-house verification requirements, provides adequate resources, and assigns trained personnel for verification activities.

Management provides work instructions that describe the process of assembling, verifying, and documenting product performance.

Senior Management Team develops, updates, and communicates the goals and plans of the Business Plan to the organization.

Controlled Copy, Do Not Duplicate **For Internal Use Only**

Figure 1-8. Finished Procedure, Page 2

Company Name ABC Company	Management Responsibility	
Division or Address 1234 Anywhere St.	Doc. No.	Rev. No.
Division or Address Anyplace, USA	Date:	Page 3 of 5

The Quality Manager conducts customer satisfaction measurements, presents the information to Senior Management, and compares the information to internal or external benchmarks.

4. PROCEDURE

4.1 Quality Policy

Senior Management Team is responsible for defining the quality policy. The quality policy specifies organizational goals and customer expectations.

Senior Management Team is responsible for implementing the quality policy. Through employee meetings, management communicates the policy to employees.

4.2 Quality Objectives

Senior Management Team defines the quality objectives. Through employee meetings and documented procedures, objectives are communicated to employees.

4.3 Management Responsibilities

Plant Manager provides necessary resources to maintain the system.

The Quality Manager conducts Management Reviews, initiates and supervises the quality system, establishes and maintains the quality system, audits implementation of the quality system, and initiates requests for, and follow-up on, corrective action.

Purchasing Manager carries out contract and order reviews.

Materials Manager selects qualified suppliers and subcontractors, assesses supplier performance, prepares and approves purchasing documents, administers storage areas, and verifies the quantity of goods received.

Operations Manager determines production personnel and equipment requirements, and ensures that processes are in control.

Production Operators perform inspection and testing in accordance with quality plans.

4.4 Management Review

See Internal Quality Audits , Corrective and Preventive Action, and quality records .

Management reviews the results of internal audits. Quality Manager plans and schedules audits of quality processes and procedures according to internal quality audits.

Semi-annually, the Senior Management Team reviews all elements of the quality system, including sector-specific and applicable customer-specific requirements.

Minutes of the Management Review are kept by the Quality Manager for a minimum of three years.

Controlled Copy, Do Not Duplicate **For Internal Use Only**

Figure 1-9. Finished Procedure, Page 3

Company Name ABC Company Division or Address 1234 Anywhere St. Division or Address Anyplace, USA	**Management Responsibility**	
	Doc. No.	Rev. No.
	Date:	Page 4 of 5

4.5 Responsibility and Authority

Job descriptions describe responsibilities and skills required by the employee and are maintained by the Human Resources Department.

Any employee may participate in a team that generates and implements procedures and work instructions.

Every employee has the authority to prevent nonconforming product.

4.6 Verification Resources

See Quality Records, xxx and Training, xxx.

Quality Manager defines where in the process testing and inspection occur. The Quality Manager and supervisors train operators to test or inspect their work.

Training records for inspection and test are maintained to verify training. Human ResourcesManager is responsible for these records. The employee training records are stored in the Human Resources Department.

Audits are performed by employees not having direct responsibility for the area being audited. Training records for employees performing audits of the quality system are maintained to verify training. Human Resources Manager is responsible for these records. The employee training records are stored in Human Resources Department.

4.7 Management Representative

Quality Manager is designated as the company's Management Representative of Quality.

The Management Representative is responsible for implementing and maintaining a quality system that meets QS 9000 requirements.

4.8 Business Plan

Senior Management Team is responsible for the preparation and revision of the ABC Business Plan. The Business Plan contains short-term and long-term goals for maintaining customer satisfaction.

The Business Plan is reviewed by Senior Management Team and updated by Purchasing Manager. Review and update of the Business Plan is done, at a minimum, yearly.

Customer expectations, trends, industry benchmarks, and world-class benchmarks are used to develop, update, and revise the Business Plan.

The Business Plan is maintained by Purchasing Manager and may be reviewed with customers by Purchasing Manager on an individual basis.

The President, or Senior Management Team, updates the organization on the Business Plan yearly. The update is accomplished by:

Controlled Copy, Do Not Duplicate **For Internal Use Only**

Figure 1-10. Finished Procedure, Page 4

Company Name ABC Company Division or Address 1234 Anywhere St. Division or Address Anyplace, USA	**Management Responsibility**	
	Doc. No.	Rev. No.
	Date:	Page 5 of 5

- Employee meetings
- Employee newsletter

5. RELATED DOCUMENTS

record of management review
employee training records
ABC Business Plan

Corrective and Preventive Action, xxx
Internal Quality Audits, xxx
Quality Records, xxx
Training, xxx

list of work instructions

Controlled Copy, Do Not Duplicate

For Internal Use Only

Figure 1-11. Finished Procedure, Page 5

Planning Your QS 9000 System and Documentation

2

What's in This Section?

This section:

- explains the types of QS 9000 documents
- provides a strategy for:
  - determining which procedure to write
  - selecting functional teams
  - gathering the relevant information

Understanding the Document Structure

Section 4.2 of the QS 9000 requires that you document your quality system and quality plans. The QS 9000 stipulates that you structure your documents according to the following four-level model:

Level A	Quality Policy Manual
Level B	Quality Procedures
Level C	Work Instructions
Level D	Quality Records

The **QS 9000 Documentation Toolkit** provides the tools to write Level B Quality Procedures.

The levels of documentation are best illustrated as a document pyramid. Refer to Figure 2-1 below. As you go higher in the pyramid, the scope widens; as you go lower in the pyramid, the detail and volume increase.

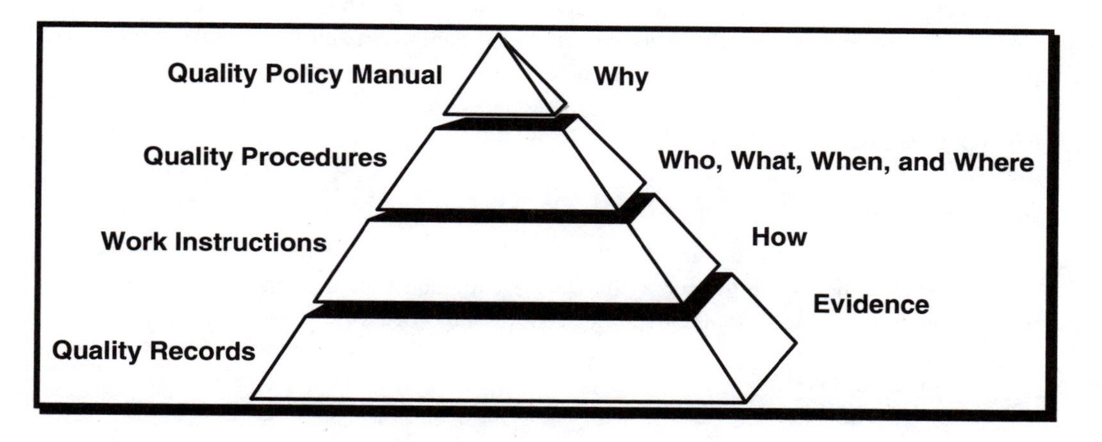

Figure 2-1. Document Pyramid

Types of Documents

This section discusses the four types of documents in the document pyramid. The **QS 9000 Documentation Toolkit** provides the tools for writing Level B Quality Procedures. Where appropriate, the worksheets offer guidance for work instructions and quality records.

The Quality Policy Manual Answers WHY

The Quality Policy Manual states your philosophy. The Quality Policy Manual should not contain proprietary information. It should address your company's philosophy for each element. The Quality Policy Manual discusses, in general terms, how you comply with each element of the QS 9000, states your quality policy, and may include your quality objectives and an overview of your company's processes. The Quality Policy Manual follows changes in customer demands and management philosophy, maintains consistency during personnel changes, reflects changes in processes, identifies related quality procedures, and is revised when new products are developed.

The ISO 10013 standard offers guidelines for writing a Quality Policy Manual. The guideline suggests including, where applicable, the scope, table of contents, introductory pages that provide general information about your organization and the quality manual itself, the quality policy and objectives, a description of the high-level structure of your organization, a description of the applicable elements, definitions, an index, and an appendix of supportive data.

Quality Procedures Answer WHO, WHAT, WHEN, and WHERE

Quality Procedures document your quality plan and define your implementation strategy. They indicate compliance to the QS 9000, demonstrate the process flow, and ensure there are no holes in the system. Quality Procedures are process-oriented and cover:

- the task: WHAT
- the responsibility: WHO
- the frequency: WHEN
- the department: WHERE

Quality Procedures are usually written at the functional department level and describe how each organization will implement their relevant part of the quality system.

Work Instructions Answer HOW

Work Instructions and documents are the step-by-step instructions specific to a product, machine, or task, and compose the bulk of the QS 9000 documentation. Work Instructions specify tools, show workmanship criteria, and indicate tolerances. Work Instructions include directions for manufacturing, packaging, receiving, customer service, purchasing, inspections and tests, and calibrations. Work Instructions may take the form of forms, reports, visual aids, and detailed instructions.

Quality Records Provide EVIDENCE

Quality Records are ongoing, objective evidence of your system and evolve from your processes. You already have many Quality Records, and you may need to create a few new ones to meet the requirements. The QS 9000 does not specify exactly which records you must keep; that is for you to define. However, most elements of QS 9000 require Quality Records. Section 4.16 of the QS 9000 directs you to store, identify, maintain, and retrieve Quality Records. You must also specify the record's retention period and assure that it complies with the retention requirements of the QS 9000.

Determining Which Procedure to Write

Before you begin to write the Level B procedures, you must determine which procedures you will write and define their scope. Typically, companies prepare between 20 and 40 procedures, though some certified companies have written more.

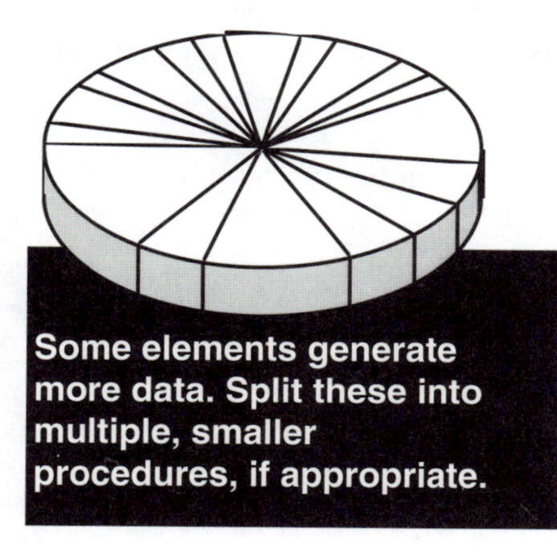

Some elements generate more data. Split these into multiple, smaller procedures, if appropriate.

It's up to you to structure your documentation. The QS 9000 describes the whole quality system. You define how you comply with the QS 9000. How you slice the whole "pie" is up to you. You can slice it into 24 sections to correspond to the 24 elements of the QS 9000. Or, you can slice it into smaller chunks, thereby increasing the number of procedures and narrowing the individual scope. What you do depends on the size and complexity of your organization. However, keep in mind that the documents must be maintained in a document control system. Weigh the quantity of documents against individual document size: more documents mean more to manage; fewer documents mean larger individual ones which may decrease usability.

To determine which procedures to write:

1. Identify the processes your facility performs.

a) Use the **QS 9000 Documentation Toolkit** to identify the scope and content of the quality procedures you will write.

Determine if the QS 9000 element has natural boundaries.

For example, the QS element Inspection and Testing may be divided into three procedures: Receiving Inspection, Incoming Inspection, and Final Inspection. Where it makes sense, separate the procedures according to the sections within a worksheet.

b) Review the forms you complete and reports you write. Typically, these are the evidence of a process.

For example, suppose you use a Corrective Action Request form. You could create a procedure that defines the activities that lead up to making the request, the activities involved in implementing the request, and the follow-up activities after closing the request.

2. Make a list of procedures.

Table 2-1 shows an example of a matrix for tracking the progress of quality procedures. Fill in the first column with procedure titles. Section 3 continues the process of completing the steps identified in the matrix.

As your documentation evolves, keep track of your progress. Appendix B provides a matrix for you to complete.

Procedure Status							
Title	Originator	Meeting: Hand out Worksheet	Meeting: Review Completed Worksheets	1st Draft Complete	Meeting: Review Comments	2nd Draft	Sign Off
Mgmt. Resp.	John Smith	dd/mm/yy	dd/mm/yy	x			
Quality System	John Smith	dd/mm/yy	dd/mm/yy	x			
Contract Review	Bill Gould	dd/mm/yy	dd/mm/yy	x	x		
Design Control	Andy Capp	dd/mm/yy	dd/mm/yy	x			
Document	Tom Milton	dd/mm/yy	dd/mm/yy	x	x		
Purchasing	Harold Wise	dd/mm/yy	dd/mm/yy	x	x		
Customer-Supplied	Charles Brown	dd/mm/yy	dd/mm/yy	x			
Product ID	Eric Sauder	dd/mm/yy	dd/mm/yy				
Process Ctrl.	Greg Jackson	dd/mm/yy	dd/mm/yy				
Inspect. & Test.	Sue Blatt	dd/mm/yy	dd/mm/yy	x	x	x	
Inspect., Meas.	Jim Marcus	dd/mm/yy	dd/mm/yy				
Test Status	Sue Blatt	dd/mm/yy	dd/mm/yy	x	x	x	
Nonconforming	Greg Jackson	dd/mm/yy	dd/mm/yy				
Corrective Action	George South	dd/mm/yy	dd/mm/yy				
H, S, P & D	Ellen Lander	dd/mm/yy	dd/mm/yy	x			
Quality Records	George South	dd/mm/yy	dd/mm/yy				
Internal Audits	George South	dd/mm/yy	dd/mm/yy				
Training	Gary White	dd/mm/yy	dd/mm/yy				
Servicing	Tom Milton	dd/mm/yy	dd/mm/yy				
Statistical Tech	Steve Burns	dd/mm/yy	dd/mm/yy	x	x	x	
PPAP	John Smith	dd/mm/yy	dd/mm;/yy	x	x	x	
Continuous Imp.	George South	dd/mm/yy	dd/mm/yy				
Manufacturing Cap.	James Brown	dd/mm/yy	dd/mm/yy				
Ford-specific	Andy Capp	dd/mm/yy	dd/mm/yy	x	x	x	
Chrysler-specific	Andy Capp	dd/mm/yy	dd/mm/yy	x			
GM-specific	Andy Capp	dd/mm/yy	dd/mm/yy	x			

Table 2-1. Procedure Status

Selecting Functional Teams

Assign an originator to lead the task of producing each procedure. Next, for each quality procedure, form a functional team composed of individuals who have access to the information. Be sure to consider that most procedures are affected by processes documented in related procedures. For example, information from your receiving inspections and your supplier corrective actions may affect how you rate suppliers for your approved vendor list. This scenario concerns implementation for the procedures on Purchasing, Inspection and Testing, and Corrective Action.

To help you in this selection process, refer to Table 2-2, which shows a model of related elements. Use this as a guideline; keep in mind that your situation may differ.

To use Table 2-2, refer to the Table Key below. The Table Key identifies Purchasing as 4.6. Now, look up 4.6 on Table 2-2. Follow across the row. There is an "X" in the columns at: 4.2, 4.3, 4.4, 4.5, 4.7, 4.8, 4.10, 4.11, 4.13, 4.14, 4.15, 4.16, 4.17, 4.18, II.1, II.2, II.3, and III. Find the procedures that correspond to these numbers in the Table Key below. You may want to elicit comments from personnel responsible for these areas since they may be affected by activities documented in the Purchasing procedure.

Key to Table 2-2:

4.1	Management Responsibility	4.13	Control of Nonconforming Product
4.2	Quality System	4.14	Corrective and Preventive Action
4.3	Contract Review	4.15	Handling, Storage, Pkg., Preserv. and Del.
4.4	Design Control	4.16	Quality Records
4.5	Document and Data Control	4.17	Internal Quality Audits
4.6	Purchasing	4.18	Training
4.7	Customer-Supplied Product	4.19	Servicing
4.8	Product ID and Traceability	4.20	Statistical Techniques
4.9	Process Control	II.1	Production Part Approval
4.10	Inspection and Testing	II.2	Continuous Improvement
4.11	Inspection, Measuring, and Test Equip.	II.3	Manufacturing Capability
4.12	Inspection and Test Status	III	Customer-Specific

	4.1	4.2	4.3	4.4	4.5	4.6	4.7	4.8	4.9	4.10	4.11	4.12	4.13	4.14	4.15	4.16	4.17	4.18	4.19	4.20	II.1	II.2	II.3	III
4.1		X			X											X	X	X				X	X	
4.2	X		X	X	X	X	X	X	X	X	X	X	X	X	X	X	X	X	X	X	X	X	X	X
4.3		X		X	X	X	X									X	X	X	X	X	X	X	X	
4.4		X	X		X	X				X						X	X	X	X		X	X		
4.5		X	X	X		X			X	X				X		X	X	X	X	X	X	X	X	
4.6		X			X				X	X				X	X	X	X	X			X	X	X	
4.7		X	X			X								X	X	X	X	X				X	X	
4.8		X	X		X	X			X					X	X	X	X	X			X	X	X	
4.9		X	X	X	X			X						X		X	X	X	X	X	X	X		
4.10		X	X	X	X	X	X	X	X		X	X	X	X	X	X	X	X	X		X	X		
4.11		X	X	X		X	X	X	X			X		X	X	X	X	X	X		X	X		
4.12		X	X				X	X	X		X			X	X	X	X	X	X		X	X	X	
4.13		X	X			X	X	X	X	X	X	X		X	X	X	X	X	X		X	X		
4.14		X	X		X	X		X	X	X	X	X			X	X	X	X	X	X	X	X		
4.15		X	X	X		X	X	X	X	X	X	X		X		X	X	X		X	X	X	X	
4.16	X	X	X	X	X	X	X	X	X	X	X	X	X	X	X		X	X	X	X	X	X	X	X
4.17	X	X	X	X	X	X	X	X	X	X	X	X	X	X	X	X		X	X	X	X	X	X	X
4.18	X	X	X	X	X	X	X	X	X	X	X	X	X	X	X	X	X		X	X	X	X		X
4.19		X	X	X	X				X	X	X	X	X	X		X	X	X		X		X		
4.20		X	X		X				X	X	X			X	X	X	X	X	X		X	X		X
II.1		X	X	X	X	X		X	X	X	X	X	X	X	X	X	X	X		X		X	X	
II.2	X	X	X	X	X	X	X	X	X	X	X	X	X	X	X	X	X	X	X	X	X		X	X
II.3	X	X	X	X		X	X	X	X		X			X		X	X	X			X	X		
III		X	X	X	X	X	X	X	X	X	X	X		X	X	X	X	X		X	X	X		

Table 2-2. Related Procedures

Gathering the Relevant Information

Now that you've selected the functional teams, the next step is to gather information.

1. Distribute a copy of the relevant worksheet to team members.

2. Members review the worksheet prior to the meeting.

3. Meet to review and collate information.

   Bring to the meeting:

   - the worksheet (completed if possible)
   - related documentation
   - relevant forms
   - flow charts of the process

4. Complete the worksheet and agree on what to include.

   When there is conflicting or missing information, resolve how you plan to implement those parts of the QS 9000.

Keep in mind that these are living documents and can be revised as you modify your system. In fact, revised documents indicate that a dynamic system is in place.

Writing the Procedures

3

What's in This Section?

This section explains:

- how to write the procedures
- how to use the Procedure Templates
- how to use the Master Template
- how to customize the formats

The Writing Process

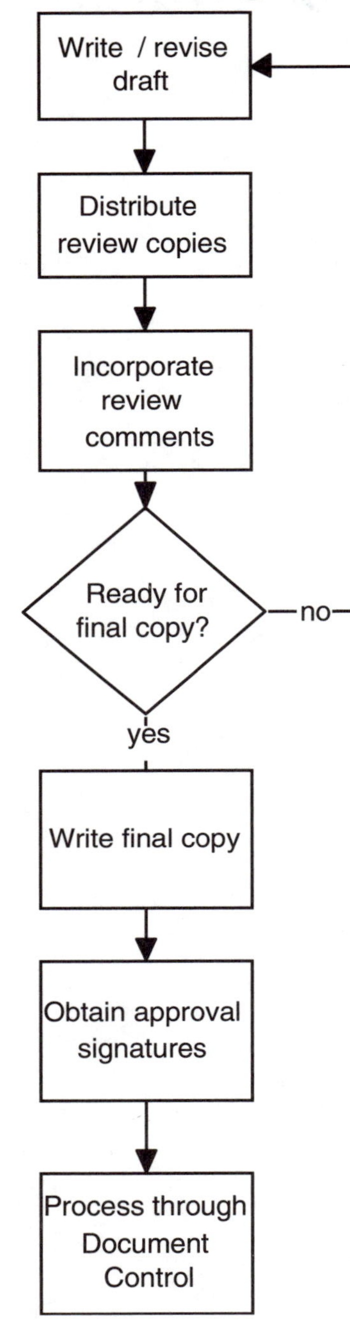

You may want to follow the steps below for writing and reviewing procedures:

1. **Write a draft of the procedure, incorporating the collected information.**

   Write so your reader can understand what you say. Keep it simple and clear. Refer to other procedures in your company for style.

2. **Distribute the draft to functional team members for comments.**

   Provide guidelines for the review. Include a cover letter specifying a date by which the review is due and any particular issues and questions for reviewers to address.

3. **Meet with the reviewers to determine which comments to incorporate into the procedure.**

   This meeting provides another opportunity to define and understand the details of the quality system you are designing.

4. **Determine if the procedure is ready for final draft.**

   If the number and type of changes to the first draft are small, incorporate these changes and write a final draft. If changes require a major rewrite of the procedure, it is best to prepare a second draft for review. Repeat 1–4.

5. **Write the final draft.**

   Incorporate changes into the final draft.

6. **Obtain approval signatures.**

   Your approval list should include the originator and those with authority to implement what is in the procedure. Revisions to the procedures must be approved by the same people on the original approval list or their designees.

7. **Process the final procedure through your document control system.**
   - Distribute approved procedures according to your distribution list.
   - Remove obsolete procedures.
   - Add this procedure to your master list of procedures.

Using the Templates

This book includes templates that can be edited on your word processor. Twenty-three procedure templates, derived from the worksheets, and one master template are provided.

See Appendix A for information about installing the software, the directory and file structure, and how to open files.

You do not need to use the master template ...

if you've determined that your procedures will correspond to the 23 elements and you are writing one procedure per element. Refer to the section "Using the Procedure Templates" for more information.

Use the master template ...

if you've determined that you will create procedures whose scope and content differ from those of the worksheets. Use the master template to create additional procedures. Refer to the sections "Using the Procedure Templates" and "Using the Master Template" for more information.

To create your own template, see the section "Customizing the Format."

Using the Procedure Templates

This section offers tips for using the templates and describes word processing commands used to customize the templates.

- Use the SAVE AS function on your word processor and rename the procedure, as appropriate.
- Always work on the renamed file. This keeps the original intact for later use.
- Always run the spell checker upon completion.
- Always back up your files.
- The procedures use Headers and Footers with the data required for document control. When you are finished writing your document, be sure to change the page count in the Header to reflect the correct number of pages.

Headers and Footers appear on every page of the procedure. See your word processor manual for information about creating and modifying Headers and Footers. Embedded in the Headers and Footers are elements required for Document Control. Pages are numbered page X of Y, where X is the page number and Y is the total number of pages.

Figure 3-1 illustrates a Header and Figure 3-2 shows a Footer.

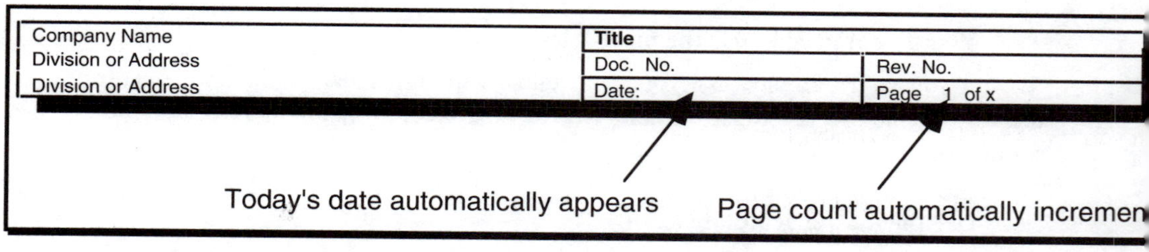

Company Name Division or Address Division or Address	**Title**	
	Doc. No.	Rev. No.
	Date:	Page 1 of x

Today's date automatically appears

Page count automatically incremen

Figure 3-1. Procedure Template Header

Controlled Copy, Do Not Duplicate

For Internal Use Only

Figure 3-2. Procedure Template Footer

- The procedures use style sheets.

  This method provides consistency. You can change a paragraph style throughout the document by modifying the style sheet. Table 3-1 lists the customized styles and specifications embedded in the templates. See your word processor manual for information about creating and modifying styles.

Style	Measurements
Approval Line	12 pt; left margin = 0, tab = 1.75
Hanging Indent 1	12 pt, left margin = 0, indent = 0.5
Hanging Indent 2	12 pt, left margin = 0.5, indent = 1
Hanging Indent 3	12 pt, left margin = 1, indent = 1.5
Hanging Indent 4	12 pt, left margin = 1.5, indent = 2
Header 1	12 pt, bold, left margin = 0
Header 2	12 pt, bold, left margin = 0.5
Paragraph 1	12 pt, left margin = 0
Paragraph 2	12 pt, left margin = 0.5
Paragraph 3	12 pt, left margin = 1
Paragraph 4	12 pt, left margin = 1.5
Related Doc	12 pt, left margin = 0.5, tab = 1.25
Table Text	12 pt
Title	14 pt, bold, left margin = 0

Table 3-1. Customized Styles

Using the Master Template

We've provided a master template with the layout designed for easy reading. If you have a layout or format you are currently using, see the section "Customizing the Format."

The master template allows you to create additional procedures. The master template incorporates the style sheet, Headers and Footers, and Section Heads. There is no content for you to edit in the master. Simply copy the desired content from the procedure template into the master. Figure 3-3 shows the first page of the master template. Figure 3-4 shows the second page.

Use the master template to:

- split a procedure template into several documents
- create a new procedure

To create a procedure from the master template:

1. Open the master template.

2. Select the SAVE AS function and name the new procedure.

This lets you make a working copy of the master template. Use the working copy to produce your procedures.

3. Open the procedure template whose contents you want to move to another document.

Skip to Step 5 if you are creating a new procedure.

Refer to the section "Using the Procedure Templates" for more information.

4. Copy the desired text from the procedure template and paste it into the master template.

When you copy the text, the paragraph styles move to the working copy.

Refer to your word processing manual for copy/paste techniques.

5. Enter or edit the text for your procedure.

6. Be sure to edit the Header and Footer, as necessary.

Refer to your word processing manual for information on Headers and Footers.

7. Run the spell checker.

8. Make a backup copy onto a diskette.

Store the diskette in a safe place.

Company Name	Title	
Division or Address	Doc. No.	Rev. No.
Division or Address	Date:	Page 1 of x

your logo

QS Procedure

Title

Approved title	Name
Approved title	Name
Approved title	Name
Approved title	Name
Approved title	Name

Change Record

Rev	Date	Responsible Person	Description of Change
A	Date	Name	Initial Release

Distribution List

(list the departments that receive controlled copies)

Controlled Copy, Do Not Duplicate   For Internal Use Only

Figure 3-3. First Page, Master Template

Company Name	Title	
Division or Address	Doc. No.	Rev. No.
Division or Address	Date:	Page 2 of x

1. Purpose

2. Scope

3. Responsibilities

4. Procedure

4.1

5. Related Documents

Copy and paste text from procedure template or write your own text.

Controlled Copy, Do Not Duplicate **For Internal Use Only**

Figure 3-4. Second Page, Master Template

Customizing the Format

If you already have a format you are using for your procedures, you can still use the contents of the procedure templates.

First, you will need to create a customized master template. Then, you must copy the contents of the procedure template to your customized master. Refer to your word processor manual for more information.

To create a customized master template:

1. Open a document that you already have which uses the desired format.
2. Select the SAVE AS function and rename the file as your master template.
3. Delete any text you do not need.
4. Edit this master to include:
   - heading styles
   - paragraph styles
   - Headers and Footers
5. Select the SAVE function.

You now have a master template to use to create the remainder of your procedures.

Notes:

Management Responsibility

4.1

What is the job title and name of the person responsible for this procedure?

ISO 9000 Standard:

4.1 Management responsibility

4.1.1 Quality policy

The supplier's management with executive responsibility shall define and document its policy for quality, including objectives for quality and its commitment to quality. The quality policy shall be relevant to the supplier's organizational goals and the expectations and needs of its customers. The supplier shall ensure that this policy is understood, implemented, and maintained at all levels of the organization.

4.1.2 Organization

4.1.2.1 Responsibility and authority

The responsibility, authority, and the interrelation of personnel who manage, perform, and verify work affecting quality shall be defined and documented, particularly for personnel who need the organizational freedom and authority to:

- *a) initiate action to prevent the occurrence of any nonconformities relating to product, process, and quality system;*
- *b) identify and record any problems relating to the product, process, and quality system;*
- *c) initiate, recommend, or provide solutions through designated channels;*
- *d) verify the implementation of solutions;*
- *e) control further processing, delivery, or installation of nonconforming product until the deficiency or unsatisfactory condition has been corrected.*

4.1.2.2 Resources

The supplier shall identify resource requirements and provide adequate resources, including the assignment of trained personnel (see 4.18), for management, performance of work, and verification activities, including internal quality audits.

4.1.2.3 Management representative

The supplier's management with executive responsibility shall appoint a member of the supplier's own management who, irrespective of other responsibilities, shall have defined authority for:

a) ensuring that a quality system is established, implemented, and maintained in accordance with this American National Standard, and
b) reporting on the performance of the quality system to the supplier's management for review and as a basis for improvement of the quality system.

NOTE 5 The responsibility of a management representative may also include liaison with external parties on matters relating to the supplier's quality system.

4.1.3 Management review

The supplier's management with executive responsibility shall review the quality system at defined intervals sufficient to ensure its continuing suitability and effectiveness in satisfying the requirements of this American National Standard and the supplier's stated quality policy and objectives (see 4.1.1). Records of such reviews shall be maintained (see 4.16).

QS 9000 Interpretations and Supplemental Quality System Requirements

The QS 9000 supplements to ISO 9001, 4.1, "Management Responsibility," are:

- Organizational Interfaces
- Management Review
- Business Plan
- Analysis and Use of Company-Level Data
- Customer Satisfaction

These additions require the supplier to analyze the entire quality system, conduct strategic business planning, and benchmark performance in relation to competitors' performance.

Suggested Procedures

QS 9000:

- Customer Satisfaction Indicators/Quality Operating System
- Customer Satisfaction Survey
- Business Plan

Quality Policy

Management with executive responsibility for quality must define the quality policy and objectives.

Does the quality policy specify organizational goals and customer's expectations? ❒*yes* ❒ *no*

Who (job title, departments) is responsible for implementing the quality policy?

Describe how Management communicates the policy to employees (i.e., during training of new hires, training for new positions, QS awareness workshops for staff, training certifications, through employee bonuses based on the quality of one's work, by posting quality statements throughout the facility, on-line notifications).

Quality Objectives

Who (job title, department managers) defines the quality objectives?

How are the objectives communicated to employees (i.e., the individual QS procedures state the pertinent quality policy or objective, they are stated in work instructions, they are taught in training)?

Management Responsibility

Below is a list of management functions. The following page also lists suggested functions. Next to each function, indicate the responsible department. Add to the list of suggested departments as it applies. Fill in only the functions that apply to your company.

Suggested Departments:

General Manager	*Sales and Marketing*	*Production*
Operations	*Purchasing*	*Quality*
Planning/ Scheduling	*Design*	*Distribution*
Engineering	*Document Control*	*Service*
Calibration	*Plant Manager*	*Warehouse*

Job Functions:

Provides resources necessary to maintain system ______________________
Prepares and updates Business Plan ______________________
Conducts audits of quality system ______________________
Analyzes market to establish desired quality characteristics of products ______________________
Establishes quality of products and services ______________________
Advertises and promotes products emphasizing quality aspects ______________________
Monitors the quality of competitors ______________________
Carries out contract and order reviews ______________________
Provides customer liaison and service ______________________
Handles customer complaints ______________________
Selects qualified suppliers and subcontractors ______________________
Prepares and approves purchasing documents ______________________
Verifies quality and quantity of received goods ______________________
Monitors and assesses supplier performance ______________________
Leads APQP efforts in accordance with the APQP and Control Plan Reference Manual ______________________
QS 9000: Prepares product specifications ______________________
QS 9000: Designs products ______________________
QS 9000: Initiates design reviews ______________________
QS 9000: Verifies and tests designs ______________________
QS 9000: Documents design outputs ______________________
Prepares control plans ______________________
Monitors control plans ______________________
Performs production engineering ______________________
Prepares production plans ______________________
Determines production personnel and equipment requirements ______________________
Conducts manufacturing planning ______________________
Conducts capacity planning ______________________
Controls and monitors processes ______________________
Defines workmanship standards ______________________
Maintains production equipment ______________________
Administrates storage areas ______________________
Establishes and maintains quality management system ______________________
Audits implementation of quality system ______________________
Initiates requests for and follow-up on corrective actions ______________________
Maintains and calibrates measuring and test equipment ______________________
Carries out supplier quality surveys and audits ______________________
Performs inspections and testing in accordance with quality plans ______________________

Handles nonconforming products ______________________
Coordinates document control activities ______________________
Maintains inspection records ______________________
Determines training requirements ______________________
Maintains training records ______________________
Tracks and determines customer satisfaction ______________________
Leads DFMEA preparation and update ______________________
Leads PFMEA preparation and update ______________________
Conducts feasibility reviews ______________________
Conducts process capability studies ______________________
Prepares PPAP submittals and coordinates changes to process or product ______________________
Administers structured problem-solving methods ______________________
Maintains shipping and packaging criteria ______________________
Maintains 100% on-time shipping notification system ______________________
Administers servicing program ______________________
Leads selection of appropriate statistical tools ______________________
Develops action plans for continuous improvement ______________________
Identifies candidates for continuous improvements actions ______________________
Leads mistake-proofing program ______________________
Maintains tool design and fabrication program ______________________
Assures customer-specific quality requirements are met ______________________

Management Review

General

Management must review the quality system for effectiveness and suitability.

Does Management review the results of internal audits? ❐ *yes* ❐ *no*

Does Management review the results of corrective actions? ❐ *yes* ❐ *no*

Does Management initiate preventive action? ❐ *yes* ❐ *no*

If yes, describe the process.

As a result of the review, are measurements added or deleted as necessary? ❐ *yes* ❐ *no*

Review Team

Management with executive responsibility for quality must review the quality system in accordance with ISO 9000/QS 9000 and your quality policy.

Is there a review team that reviews all elements of the quality system, including sector- and customer-specific requirements? ❐ *yes* ❐ *no*

If yes, complete the following:

What is the name of the team?

Who (job titles) comprise the team?

How frequently does the team review the corporate quality measures?

How frequently does the team meet?

Records

Are records of the review meetings maintained? ❐ *yes* ❐ *no*

If yes, complete the following:

What is the title of the record (i.e., meeting minutes)?

Where are the records stored?

How long are the records kept?

You must keep job descriptions for personnel whose work affects quality.

Responsibility and Authority

Job Descriptions

Do job descriptions describe responsibilities and skills required by the employee? ❒yes ❒ no

If yes, where are they maintained (i.e., within individual departments, Human Resources)?

Authority

Who (job title, i.e., Department Managers) generates and implements procedures and work instructions?

Who (job title, all employees) has the authority to prevent nonconforming product?

Who (job title, responsible department) maintains quality records?

Quality Problems and Solutions

Who (job title) has authority to stop production or delivery until deficiencies are corrected?

Who (any employee, job title) can initiate solutions to a quality-related problem?

What is the procedure (i.e., contact Supervisor, Group Leader, or Quality Control, complete a form)?

How is the success of a solution to a quality problem verified (i.e., the corrective action process)?

Verification Resources

Inspection and Test

You must identify what needs verification, provide the resources, and train personnel, which includes inspection, testing, and monitoring of design, production, installation, and services.

Where in the process do inspections and tests occur (i.e., design verification, in-process inspections, final inspections, installation, service)?

Who (job title) defines where in the process testing and inspection occur?

Who (job title) trains the operators to test or inspect their work?

Are records of training for inspection and testing maintained? ❒yes ❒ no

If yes, complete the following:

What is the title of the training record or form verifying training?

Who (job title) is responsible for these records?

Where are these records stored?

Audit Training

Audits must be performed by personnel not having direct responsibility for the work.

Do internal auditors audit areas other than the ones for which they are responsible? ❐*yes* ❐ *no*

What training is provided to employees performing audits of the quality system (i.e., attendance in workshops in internal auditing)?

Are records of audit training kept? ❐*yes* ❐ *no*

If yes, complete the following:

What is the title of the audit training record or form?

Who (job title) is responsible for the audit training records?

Where are these records stored?

Design Review Training

Design reviews must be verified by personnel not having direct responsibility for the work.

Do reviewers work in areas other than the ones for which they are reviewing? ❐*yes* ❐ *no*

What training is provided to reviewers who verify design?

Are records of design review training kept? ❐*yes* ❐ *no*

If yes, complete the following:

What is the title of the training record or form?

Who (job title) is responsible for the design review training records?

Where are these records stored?

Management Representative

Who (job title and person's name) is designated as the company's Management Representative of Quality?

The management representative has executive responsibility for quality and must be a member of management.

Does the management representative have the authority to review that the quality system is implemented and maintained in accordance with QS 9000? ❒*yes* ❒ *no*

Does the management representative report directly to executive management? ❒*yes* ❒ *no*

Organizational and Technical Interfaces

Who (job title) is responsible for managing activities during:

- Concept development
- Prototype
- Production

Activities during concept development, prototype, and production must be properly managed. These activities are described in the APQP and Control Plan Reference Manual.

Do the activities during each of these phases meet the intent of the Advanced Product Quality Planning and Control Plan Reference Manual? ❒yes ❒ no

You must use a multi-disciplinary approach to decision making.

Do cross-functional teams make decisions regarding product and process during these phases? ❒yes ❒ no

Are communication channels and appropriate methods of communication established with the customer? ❒yes ❒ no

If yes, can data be provided in a format acceptable to the customer? ❒yes ❒ no

Business Plan

You must have a formal, documented, comprehensive Business Plan that covers short-term (1-2 years) and long-term (over 3 years) goals.

Is there a formal, documented, comprehensive Business Plan for your facility? ❒yes ❒ no

If yes, complete the following:

Who (job title/team name) is responsible for preparing the document?

Who (job title) is responsible for maintaining the document?

Where is the document located?

How often is the Business Plan updated (i.e., semi-annually, annually, significant market change)?

Who (job title/team name) is responsible for updating the Business Plan?

Who (job title) is responsible for revising the Business Plan?

Who (job title/team name) is responsible for reviewing the Business Plan?

Who (job title) is responsible for approving the Business Plan?

Who (job title) is responsible for communicating the information to the appropriate members of the organization?

How is the information in the Business Plan communicated to the organization (i.e., state-of-the-company meetings, employee fact sheets, department meetings)?

Are the business goals and plans based on:

analysis of competitive product? ❒yes ❒ no

benchmark data for best-in-class and world-class? ❒yes ❒ no

results of customer surveys ❒yes ❒ no

If no to the above question, what are the goals and plans based on?

You must use a proven process for collecting objective data to determine the customer's current and future expectations.

How are the data for goals and plans identified?

What methods are used to determine customers' current expectations (i.e., customer surveys, design reviews, market information, benchmarking of key product or service features)?

What is the frequency and method for collecting this information (i.e., annual customer surveys, quarterly key customer meeting)?

What methods are used to determine customers' future expectations (i.e., customer surveys, design reviews, market information, benchmarking of key product or service features, R&D reviews)?

What is the frequency and method for collecting this information (i.e., yearly strategic business sessions with customers, program reviews)?

Complete the following matrix. Indicate with a yes or no whether each topic is addressed in the Business Plan for both the long- and short-terms. In the third column, indicate who (job title) is responsible for implementing each topic of the Business Plan.

Topic	Short-term	Long-term	Responsible
Market-related issues			
Financial planning and cost			
Growth projections			
Plant/facilities plans			
Distribution plans			
Cost objectives			
Human resource development			
R&D plans, projections, funding			
Projected sales figures			
Marketing plans			
Quality objectives			
Customer satisfaction plans			
Key internal quality performance measurables			
Key internal operation performance measurables			
Health, safety, and environmental issues			

Analysis and Use of Company Level Data

Is there a documented procedure for analyzing trends in the operational and quality performance of key product and service features? ❒yes ❒ no

How are the key product and service features selected?

You must track and document trends in the quality and operational performance of key product and service features. You must also document the current quality levels of these features.

Do you have an existing Quality Operating System (QOS)? ❒yes ❒ no

If yes, do the QOS measurables support the overall business goals of the organization? ❒yes ❒ no

How is the QOS information used in developing/updating/revising the Business Plan?

What quality and operational performance (i.e., productivity, efficiency, effectiveness) indicators are tracked?

Is this information compared with competitors or other appropriate benchmarks? ❒yes ❒ no

Is corrective action taken if there is an unfavorable trend or result? ❒yes ❒ no

Is this information used for assessing overall business objectives? ❒yes ❒ no

Who (job title) is responsible for maintaining this information?

Trends and current performance indicators should provide information for meeting overall business goals. Benchmarking practices should be used in determining the goals.

Is the information translated into action items (i.e., adverse trends and/or customer concerns assigned to teams for corrective action)? ❒yes ❒ no

Does this information influence company decision-making and long range planning? ❒yes ❒ no

Customer Satisfaction

You must have a documented process for assessing customer satisfaction and readily identifying customer dissatisfaction.

Is there a documented procedure for determining customer satisfaction? ❒yes ❒ no

If yes, complete the following:

What is the frequency for determining customer satisfaction (i.e., quarterly, semi-annually)?

Senior management must review customer satisfaction benchmarks and trends.

Who (job title, department managers, Senior Management Team) reviews customer satisfaction information?

How is the validity and objectivity of the information and conclusions attained?

What documentation and supporting information (i.e., trend charts, Paynter charts) of internal and external customer satisfaction are maintained?

Is this information part of the Analysis of Company Level Data or QOS procedure?

How is internal and external customer dissatisfaction determined?

Who (job title) is responsible for maintaining the documentation?

How is this information compared with competitor performance or appropriate benchmarks?

Written Procedures and Related Records

List records or forms maintained when processing activities described in this procedure:

List related work instructions and procedures with their corresponding part numbers that employees use as instructions for performing activities described in this procedure:

MANAGEMENT RESPONSIBILITY PROCEDURE TEMPLATE

1. PURPOSE

- ***To establish a mechanism for management to set quality goals and review their attainment.***
- ***To provide for management review of internal quality audits and completion of corrective action.***
- ***To ensure all employees are adequately trained.***
- ***To establish and implement a quality system by formulating the quality policy, defining organization, assigning authorities and responsibilities, appointing the management representative, reviewing the quality system, and ensuring the resources and personnel to maintain the system.***
- ***To clearly identify the management function responsible for the Quality System.***
- ***To establish a quality policy and objectives and ensure that the policy is communicated and practiced throughout the organization and that quality objectives are routinely met by everyone.***
- ***To have in place an organization chart and a description of roles and responsibilities required for the business unit. The organization chart is a controlled document and updated at least once a year.***
- **To establish and implement a system for strategic planning through the use of a documented Business Plan. The Business Plan includes the analysis of company-level data, benchmarked information, and customer satisfaction indicators.**

2. SCOPE

This procedure applies to all operations at <Company>.

3. RESPONSIBILITIES

The <job title> has overall responsibility for quality at <Company>.

The QS 9000 Team, composed of the <job titles>, is responsible for forming an internal audit team to audit QS 9000-compliance.

The Senior Management Team, composed of <job titles>, analyzes company-level data and customer satisfaction indicators.

The <job title> prepares, updates, and revises the Business Plan.

The <job title> approves the Business Plan.

The <job title, Team Name> reviews the Business Plan.

Management supports the organizational freedom and authority to:

- ***Identify and record any product quality problems.***
- ***Initiate, recommend, or provide solutions through designated channels.***
- ***Verify the implementation of solutions.***
- ***Control further processing, delivery, or installation of nonconforming product until the deficiency or unsatisfactory condition has been corrected.***

Management defines the responsibility, authority, and interrelation of all personnel who manage, perform, and verify work affecting quality. Management identifies in-house verification requirements, provides adequate resources, and assigns trained personnel for verification activities.

Management provides Work Instructions that describe the process of assembling, verifying, and documenting product performance.

<Job titles or Team Name> develops, updates, and communicates the goals and plans of the Business Plan to the organization.

<Job title> conducts customer satisfaction measurements, presents the information to Senior Management, and compares the information to internal or external benchmarks.

4. PROCEDURE

4.1 Quality Policy

<Job title> is responsible for defining the quality policy. The quality policy specifies organizational goals and customer expectations.

<Job title> is responsible for implementing the quality policy. Through <activities> management communicates the policy to employees.

4.2 Quality Objectives

<Job title> defines the quality objectives. Through <activities> objectives are communicated to employees.

4.3 Management Responsibilities

Quality <list job functions>.

Operations <list job functions>.

Planning <list job functions>.

Engineering <list job functions>.

Sales and Marketing <list job functions>.

Purchasing <list job functions>.

Document Control <list job functions>.

Production <list job functions>.

Distribution <list job functions>.

Service <list job functions>.

Plant Manager <list job functions>.

Calibration <list job functions>.

Warehouse <list job functions>.

4.4 Management Review

See *Internal Quality Audit, Corrective and Preventive Action, and Quality Records .*

Management reviews the results of internal audits. <Job title> plans and schedules audits of quality processes and procedures according to Internal Quality Audits.

<Frequency>, <Team Title>, composed of <job titles>, reviews all elements of the quality system including sector-specific and applicable customer-specific requirements.

<Frequency>, <Team Title>, composed of <job titles>, reviews corrective actions.

<Frequency>, <Team Title>, composed of <job titles>, analyzes and evaluates preventive actions.

As a result of the reviews, measurements are added or deleted as necessary.

<Name of record> is kept at the team meetings. The <name of record> is stored in <department> and maintained by <job title> for <length of time>.

4.5 Responsibility and Authority

Job descriptions describe responsibilities and skills required by the employee and are maintained by <department>.

<Job title> generates and implements procedures and work instructions.

<Job title> has the authority to prevent nonconforming product.

4.6 Quality Problems and Solutions

Any employee can initiate solutions to a quality-related problem by <activity>. The success of a solution to a quality problem is verified by <activity>.

4.7 Verification Resources

See *Quality Records, xxx and Training, xxx.*

<Job title> defines where in the process testing and inspection occur. <Job title> trains operators to test or inspect their work.

Training records for inspection and test are maintained to verify training. <Job title> is responsible for these records. The <names of training records> are stored in <department> and maintained by <job title>.

Audits are performed by employees not having direct responsibility for the area being audited. Training records for employees performing audits of the quality system are maintained to verify training. <Job title> is responsible for these records. The <names of training records> are stored in <department> and maintained by <job title>.

Training records for employees performing design reviews are maintained to verify training. <Job title> is responsible for these records. The <names of training records> are stored in <department> and maintained by <job title>.

4.8 Management Representative

<Job title> is designated as the company's Management Representative of Quality.

The management representative is responsible for:

- *<Activity>*
- *<Activity>*

4.9 Organizational and Technical Interfaces

See Advanced Product Quality Planning, xxx.

During each phase of product development, the following employees are responsible for managing and coordinating product development activities:

- **Concept development-<job title>**
- **Prototype-<job title>**

- Production-<job title>

Decisions regarding the product or process are made by a cross-functional team in accordance with the Advanced Product Quality Planning Procedure <insert procedure number>. The Advanced Product Quality Planning and Control Plan Reference Manual is used to identify the activities performed during each phase of development.

4.10 Analysis and Use of Company-level Data

See Corrective and Preventive Action, xxx.

The <Management Team> selects measurables that indicate the quality performance, operational performance, and current quality level of key product and service features.

The <Management Team> assigns a performance goal to each indicator that supports the overall business goals of the company. The goals are established by:

- <Activity>
- <Activity>

A member of the <Management Team> is assigned to track each of the operational or quality performance indicators and the current quality level of that item. The performance of the product or service feature is tracked by collecting data, analyzing the data and presenting the data to the <Management Team>.

The <Management Team> compares the overall performance and current quality levels to the goal <frequency>. The <Management Team> requires timely solutions to customer-related problems, adverse trends, or information that impacts overall business goals. A member of the <Management Team> is designated to lead a team that develops the solutions and presents the solution(s) to the <Management Team>.

4.11 Customer Satisfaction

See Continuous Improvement, xxx.

See Corrective and Preventive Action, xxx.

The <job title> is responsible for conducting <frequency> customer satisfaction assessments. <Job title> maintains the results of the customer assessments.

The results of the assessments are reviewed by the <Senior Management Team> to identify trends in customer satisfaction and indicators of customer dissatisfaction. Industry benchmarks are used when reviewing the customer assessments.

Pertinent information from the customer assessments are used to initiate corrective action. The information is used during the review, update, and revision of the <Company> Business Plan.

4.12 Business Plan

<Job title, Team Name> is responsible for the preparation and revision of the <Company> Business Plan. The Business Plan contains short-term and long-term goals for maintaining customer satisfaction.

The Business Plan is reviewed by <job title, Team Name> and updated by <job title>. Review and update of the Business Plan is done, at a minimum, <frequency>.

Customer expectations, trends, industry benchmarks, and world-class benchmarks are used to develop, update, and revise the Business Plan.

The Business Plan is maintained by <job title> and may be reviewed with customers by <job title> on an individual basis.

<Job title, Team Name> updates the organization on the Business Plan <frequency>. The update is accomplished by:

- **<Activity>**
- **<Activity>**

5. RELATED DOCUMENTS

<record of management review>
<training records>
<Business Plan>
<customer assessment records>
<company-level data records>
Corrective and Preventive Action, xxx

Internal Quality Audits, xxx
Quality Records, xxx
Training, xxx

Continuous Improvement, xxx

<list of work instructions>

Notes:

Quality System

4.2

What is the job title and name of the person responsible for this procedure?

ISO 9000 Standard:

4.2 Quality system

4.2.1 General

The supplier shall establish, document, and maintain a quality system as a means of ensuring that product conforms to specified requirements. The supplier shall prepare a quality manual covering the requirements of this American National Standard. The quality manual shall include or make reference to the quality system procedures and outline the structure of the documentation used in the quality system.

NOTE 6 Guidance on quality manuals is given in ISO 10013.

4.2.2 Quality system procedures

The supplier shall

a) prepare documented procedures consistent with the requirements of this American National Standard and the supplier's stated quality policy, and

b) effectively implement the quality system and its documented procedures.

For the purposes of this American National Standard, the range and detail of the procedures that form part of the quality system depend on the complexity of the work, the methods used, and the skills and training needed by personnel involved in carrying out the activity.

NOTE 7 Documented procedures may make reference to work instructions that define how an activity is performed.

4.2.3 Quality planning

The supplier shall define and document how the requirements for quality will be met. Quality planning shall be consistent with all other requirements of a supplier's quality system and shall be documented in a format to suit the supplier's method of operation. The supplier shall give consideration to the following activities, as appropriate, in meeting the specified requirements for products, projects, or contracts:

a) the preparation of quality plans;

b) *the identification and acquisition of any controls, processes, equipment (including inspection and test equipment), fixtures, resources, and skills that may be needed to achieve the required quality;*
c) *ensuring the compatibility of the design, the production process, installation, servicing, inspection, and test procedures, and the applicable documentation;*
d) *the updating, as necessary, of quality control, inspection, and testing techniques, including the development of new instrumentation;*
e) *the identification of any measurement requirement involving capability that exceeds the known state of the art, in sufficient time for the needed capability to be developed;*
f) *the identification of suitable verification at appropriate stages in the realization of product;*
g) *the clarification of standards of acceptability for all features and requirements, including those which contain a subjective element;*
h) *the identification and preparation of quality records (see 4.16).*

NOTE 8 The quality plans referred to (see 4.2.3a) may be in the form of a reference to the appropriate documented procedures that form an integral part of the supplier's quality system.

QS 9000 Interpretations and Supplemental Quality System Requirements

The QS 9000 supplements to ISO 9001, 4.2, "Quality System," are:

- Quality Planning and Special Characteristics
- Use of Cross-functional Teams
- Feasibility Reviews
- Process Failure Mode and Effects Analysis (Process FMEAs)
- The Control Plan

These additions require the supplier to perform advanced product quality planning using the methods and activities described in the Advanced Product Quality Planning (APQP) and Control Plan Reference Manual. The APQP and Control Plan Reference Manual requires the use of FMEAs as preventive measures in quality planning and the use of control plans throughout all phases of development. Suppliers must also conduct feasibility reviews to assess manufacturing feasibility prior to accepting a contract.

Suggested Procedures

QS 9000:

- Control Plans
- Failure Mode and Effects Analysis
- Advanced Product Quality Planning
- Feasibility Reviews

You must document your quality system.

Description

Is your quality system documented? ❒*yes* ❒ *no*

Are the documents controlled? ❒*yes* ❒ *no*

Quality Manual

Is the quality manual simply stated, easy to use, and fully approved by all affected functional groups? ❒*yes* ❒ *no*

Does the quality manual include references to documented procedures? ❒*yes* ❒ *no*

Is the quality manual current and up-to-date to reflect changes to the system? ❒*yes* ❒ *no*

Quality Procedures

Are the procedures and instructions:

up-to-date ❒*yes* ❒ *no*
implemented as written ❒*yes* ❒ *no*
consistent with your quality policy ❒*yes* ❒ *no*
consistent with the requirements of ISO 9000 ❒*yes* ❒ *no*
consistent with the requirements of QS 9000 ❒yes ❒ no

Quality Planning

You must have an advanced product quality planning process based on the APQP and Control Plan Reference Manual.

Is there a documented Advanced Product Quality Planning (APQP) and Control Plan Procedure that conforms to the APQP and Control Plan Reference Manual? ❒yes ❒ no

If yes, what is the title and number of that procedure?

Is the APQP and Control Plan Reference Manual used to perform quality planning for all products? ❒yes ❒ no

If no, complete the following table:

Product	(yes/no)	Explain why APQP is not used (i.e., after-market product only)

Are cross-functional teams established to perform APQP activities? ❒yes ❒ no

You must conduct manufacturing feasibility analysis prior to accepting a contract.

Are feasibility reviews conducted in accordance with APQP timelines and requirements? ❒yes ❒ no

If no, do the feasibility reviews confirm the organization's ability to manufacture the product prior to contract award? ❒yes ❒ no

Are the results of feasibility reviews documented? ❒yes ❒ no

If yes, complete the following:

What is the name of the form (i.e., Team Feasibility Commitment)?

Who (job title) maintains the form?

Special characteristics must be finalized during the preparation of FMEAs and control plans. You must obtain the required customer approvals for FMEAs and control plans.

Are Process FMEAs used to finalize special characteristics? ❒yes ❒ no

Do the Process FMEAs consider all special characteristics? ❒yes ❒ no

Are special characteristics identified per customer requirements on the Process FMEA? ❒yes ❒ no

If no, how are the special characteristics identified?

Do you obtain customer approval, if required, on FMEAs? ❒yes ❒ no

Who (job title) maintains FMEAs?

Cross-functional teams must develop control plans for each phase of product development.

Are your control plans documented? ❒yes ❒ no

Are control plans available for all affected processes? ❒yes ❒ no

Do control plans cover all phases of development, including:

- prototype ❒yes ❒ no
- pre-launch ❒yes ❒ no
- production ❒yes ❒ no

Who (job titles, department, team name) prepares the control plans?

Who (job title) approves the control plans?

You must review and update control plans when a product or process changes, or when the process becomes unstable or non-capable.

Who (job title) maintains control plans?

Who (job title) reviews and updates control plans?

Are control plans referenced in the procedures? ❒yes ❒ no

Do control plans identify the required:

development phase (i.e., prototype, pre-launch, production)	❒yes	❒ no
control plan number	❒yes	❒ no
key contact	❒yes	❒ no
origination date	❒yes	❒ no
revision date	❒yes	❒ no
part number/latest change level	❒yes	❒ no
core team	❒yes	❒ no
customer engineering approval date (if required)	❒yes	❒ no
part name/description	❒yes	❒ no
supplier/plant approval/ date	❒yes	❒ no
customer quality approval date (if required)	❒yes	❒ no
supplier/plant	❒yes	❒ no
supplier code	❒yes	❒ no
other approval date (if required)	❒yes	❒ no
part/process number	❒yes	❒ no
process name/operation description	❒yes	❒ no
machine, device, jig, tools for manufacturing	❒yes	❒ no
product characteristic	❒yes	❒ no
process characteristic	❒yes	❒ no
special characteristic classification	❒yes	❒ no
product/process specification/tolerance	❒yes	❒ no
evaluation measurement technique	❒yes	❒ no
sample size and frequency	❒yes	❒ no
control method	❒yes	❒ no
reaction plan	❒yes	❒ no

If a system other than APQP is used by the organization, complete the following questions on quality planning.

Are your quality plans documented?

Who (job title, department) prepares the quality plans?

Who (job title, department) approves the quality plans?

Are quality plans referenced in the procedures? ❒*yes* ❒ *no*

Do quality plans identify the required:

equipment	❒*yes*	❒ *no*
processes	❒*yes*	❒ *no*
fixtures	❒*yes*	❒ *no*
production resources and skills	❒*yes*	❒ *no*
measuring requirements	❒*yes*	❒ *no*
test and inspections	❒*yes*	❒ *no*
process verifications and quality controls	❒*yes*	❒ *no*
quality records	❒*yes*	❒ *no*
procedures and work instructions	❒*yes*	❒ *no*
installation and servicing requirements	❒*yes*	❒ *no*
design and production reviews	❒*yes*	❒ *no*
standards of acceptability	❒*yes*	❒ *no*

Document Structure

Does your document structure follow the figure below? ❒*yes* ❒ *no*

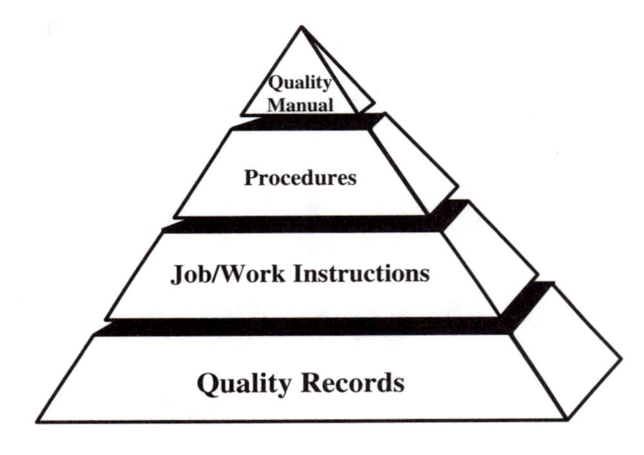

Figure 4.2-1. Document Structure

If not, describe your document structure:

Does your document control numbering system distinguish between the levels? ❒*yes* ❒ *no*

If yes, provide an example of the numbering system for each level:

QUALITY SYSTEM PROCEDURE TEMPLATE

1. PURPOSE

- *To establish and maintain a documented Quality System.*
- To establish and document a system for quality planning that uses the Advanced Product Quality Planning and Control Plan Reference Manual as the foundation for the quality plan.
- To establish a quality system that utilizes Failure Mode and Effects Analysis (FMEA) and Control Plans to confirm special characteristics during all phases of development.

2. SCOPE

The Quality System defined in this manual applies to all personnel who perform activities affecting quality.

3. RESPONSIBILITIES

All employees are responsible for the Quality System. The individual documents define specific employee responsibilities.

Any employee, if requested by <job title>, may participate on an Advanced Product Quality Product Planning Team.

<Job title/Team Name> prepares the Process Failure Mode Effects Analysis.

<Job title/Team Name> prepares and updates control plans.

<Job title/Team Name> conducts Feasibility Reviews prior to accepting a manufacturing contract.

4. PROCEDURE

4.1 Description

Through a formal documented system of planned activities, the Quality System meets:

- *Jurisdictional regulations, codes, and standards*
- *Contractual specifications and drawings*
- *Corporate quality objectives*

This program complies with the applicable sections and elements of standards established and recognized by the governmental agencies and customers served.

4.2 Quality Manual

The quality manual is current and up-to-date to reflect changes to the system. It is simply stated, easy to use, and fully approved by all affected functional groups.

<Company> defines its policy for each QS 9000 element, including sector-specific and applicable customer-specific requirements in the quality manual. *For each element, as appropriate, <Company> has documented procedures that further describe how the specific policy objectives and goals are met. The quality manual references these documented procedures.* Where applicable, work instructions are referenced in the quality manual.

4.3 Quality Procedures

Procedures and instructions are implemented as written. The procedures explain how <Company> implements the requirements of QS 9000 in accordance with its quality policy. They are revised, as necessary, to reflect the actual objectives, flow of tasks, and staff responsibilities.

Where applicable, work instructions and quality records are referenced in the documented procedures and quality manual.

4.4 Quality Plans

See APQP and Control Plan Procedure, xxx.

The <Company> prepares quality plans, as appropriate, for the products it designs and manufactures.

Advanced Product Quality Planning (APQP) is performed by cross-functional teams in accordance with the APQP and Control Plan Reference Manual for

- <Product>
- <Product>
- <Product>

The APQP and Control Plan Procedure, <procedure number>, details the implementation of APQP at <Company>.

Feasibility reviews are conducted by <job titles, Team Name> on <automotive customer products> in accordance with the APQP timelines and requirements. Records of feasibility reviews are maintained by <job title> on <name of form>.

Failure Mode Effects Analysis (FMEAs) are used to confirm special characteristics. An FMEA addresses all special characteristics initially identified by the customer and special characteristics that are confirmed through the APQP process.

FMEAs are prepared using the Potential Failure Mode and Effects Analysis Reference Manual.

FMEAs are prepared by <job title, team name> and are maintained by <job title>.

FMEAs are reviewed by <job title, team name> whenever a product or process changes.

Documented Control Plans are prepared by <job titles, team name> and address, if appropriate, all phases of development (prototype, pre-launch and production). The <job title> approves the control plans. Control Plans are maintained by <job title>.

Control Plans are prepared using the APQP and Control Plan Reference Manual.

<Job title> prepares quality plans for the following products that do not follow the APQP and Control Plan process:

- *<Product>*
- *<Product>*
- *<Product>*

Each quality plan for the referenced product, specifies the equipment and fixtures required to build the product, the resources and skills, what tests and verifications will be performed to measure process and product quality, the records and written documentation used by personnel to build or service the product, any installation and service requirements, the schedule and content of production and design reviews, and the standards of acceptability.

The quality plans are approved by <job titles, departments> and are maintained in the Document Control System.

4.5 Document Structure

LEVEL 1	*Quality Manual*
LEVEL 2	*QS 9000 Procedures Manual*
LEVEL 3	*Job Instructions*
LEVEL 4	*Quality Records*

5. Related Documents

APQP and Control Plan Reference Manual
Potential Failure Mode and Effects Analysis (FMEA) Reference Manual
FMEA Form
Control Plan Form
Feasibility Review Form
APQP and Control Plan

<list of work instructions>

Contract Review

4.3

What is the job title and name of the person responsible for this procedure?

ISO 9000 Standard:

4.3 Contract review

4.3.1 General

The supplier shall establish and maintain documented procedures for contract review and for the coordination of these activities.

4.3.2 Review

Before submission of a tender, or at the acceptance of a contract or order (statement of requirement), the tender, contract, or order shall be reviewed by the supplier to ensure that:

a) the requirements are adequately defined and documented; where no written statement of requirement is available for an order received by verbal means, the supplier shall ensure that the order requirements are agreed before their acceptance;

b) any differences between the contract or accepted order requirements and those in the tender are resolved;

c) the supplier has the capability to meet the contract or accepted order requirements.

4.3.3 Amendment to contract

The supplier shall identify how an amendment to a contract is made and correctly transferred to the functions concerned within the supplier's organization.

4.3.4 Records

Records of contract reviews shall be maintained (see 4.16).

NOTE 9 Channels for communication and interfaces with the customer's organization in these contract matters should be established.

QS 9000 Interpretations and Supplemental Quality System Requirements

The QS 9000 supplement to ISO 9001, 4.3, "Contract Review," is that the supplier must review all customer-specific requirements prior to accepting a contract or purchase order.

Suggested Procedures:

- *Generating Quotes*
- *Sales Order Entry*

QS 9000:

- Planning Production

General

For which products are quotes generated?

Are products priced in company product literature catalogs? ❒ *yes* ❒ *no*

How often are the catalogs updated?

Who (job title) is responsible for updating the catalogs?

Who receives these catalogs (i.e., general public, your sales staff, distributors)?

Who (job title/group, i.e., Sales, Request for Quotation Committee) reviews contracts?

If a group reviews contracts, who (job titles) comprise the group?

Marketing Requirements

Who (job title, i.e., Product Manager) projects the market?

What data are used for market projection (i.e., historical trends, seasonal data, promotions, programs, customer input)?

Is customer input gathered? ❐*yes* ❐ *no*

If yes, complete the following:

How is customer information collected (i.e., telephone and written correspondence, feedback from beta sites, direct mail inquiries, interviews at trade shows)?

Who (job title, i.e., Marketing, R & D Engineering) identifies features to put into future products?

Generating Quotes

Requests for Quotes

Are requests for quotes reviewed to determine whether to bid on a quote? ❐*yes* ❐ *no*

If yes, complete the following:

Who (job title, group title) is responsible for determining whether to bid on a quote?

What criteria is used to determine whether to bid (i.e., feasibility study completed, APQP Team approval)?

Who (job title) is responsible for tracking the bid?

How is the bid tracked (on a form, database, bid number assigned)?

If a form is used, complete the following:

What is the title of the form?

Who (job title) is responsible for maintaining the form?

Where is the form stored?

If a database is used, complete the following:

Who (job title) is responsible for maintaining the database?

Where is the database stored (department, plant)?

Who (job titles) have access to the database?

Compose the Quote

Who (job title) is responsible for preparing the quote?

What information is used to provide a quote (i.e., company price list, customer specifications)?

Who (job title) is responsible for estimating material and labor costs?

Is a form completed with the new quote information? ❒ *yes* ❒ *no*

If yes, complete the following:

Who (job title) completes the form?

What is the title of the form?

Are different approvals required depending upon the type of sale? ❒ *yes* ❒ *no*

If yes, describe:

Contact Customer

Who (job title) contacts the customer with an approved quote?

Is a quotation form sent to the customer? ❐*yes* ❐ *no*

If yes, complete the following:

What is the title of the quotation form?

What information is entered on the form (i.e., product description, price, delivery, relevant technical literature)?

How are quotations tracked (i.e., quote numbers)?

Order Entry

General

Are different types of orders processed differently? ❐*yes* ❐ *no*

If yes, list the types of orders (i.e., domestic, international, spares, service contracts)?

You must keep records of the sales contract.

Placing Orders

How does the customer place an order?

Are orders recorded onto a form? ❒yes ❒ no

If yes, complete the following:

What is the title of the form?

Who (job title) maintains the form?

Where is the form stored?

Verifying Orders

You must define and document your customer's require-ments.

Are customers assessed to determine if they have ordered previously? ❒yes ❒ no

Are orders verified? ❒yes ❒ no

If yes, complete the following:

What is verified (i.e., quotation on item, pricing, availability, customer information, shipping information, and appropriateness for customer's requirements)?

Who (job title) is responsible for verifying the order?

Are credit terms established? ❒yes ❒ no

If yes, who is responsible for establishing the terms?

Tracking Customers

Are customers tracked? ❐ *yes* ❐ *no*

If yes, complete the following:

Who (job title) is responsible for keeping track of the customers?

What information is tracked?

Where is the customer information recorded (database, customer file)?

Sales Orders

You must resolve requirements that differ from those tendered.

Are sales orders generated? ❐ *yes* ❐ *no*

If no, do you have a verbal agreement? ❐ *yes* ❐ *no*

If yes, complete the following:

Who (job title) prepares the sales order?

Who (job title) is responsible for sending the sales order to the customer?

Do you require customer acknowledgment of the sales order? ❐ *yes* ❐ *no*

You must store records of contract review as a quality record.

If yes, complete the following:

Where is the customer acknowledgment maintained?

Who (job title) is responsible for maintaining the acknowledgment?

Amendments to Contracts

If an amendment to the contract is made, who (job title) informs the relevant groups?

Who (job title) updates the sales order?

Ship Dates

Are ship dates provided with the sales order? ❒yes ❒ no

If yes, complete the following:

How is the ship date estimated?

Who (job title) is responsible for estimating the ship date?

Who (job title) receives the sales order (i.e., Production, Shipping)?

Foreign Sales

For foreign sales, is any other documentation required? ❒yes ❒ no

If yes, explain the requirements.

Production Control

How would you classify your operating environment (repetitive, just-in-time, MRP)?

How are production levels determined (based on machinery capacity, sales orders, forecasts)?

Draw a flow chart of the production planning process indicating the steps and responsible parties for each step of the process.

Are work orders generated? ❒ *yes* ❒ *no*

If yes, complete the following:

Who (job title) is responsible for releasing the work order?

What is included in the work order (i.e., operation record, drawings, bill of materials, setup sheet, labor hours, build quantity, routings, the work center)?

Is the work order software-generated? ❒ *yes* ❒ *no*

If yes, complete the following:

Which software program are you using, if applicable?

Do you generate a Materials Requirement Plan? ❒ *yes* ❒ *no*

Do you produce a schedule that lists the raw materials needed? ❒ *yes* ❒ *no*

Do you generate a capacity plan to assess the available hours per work center and the hours required to manufacture the subassembly or assembly? ❒ *yes* ❒ *no*

Who (job title) is responsible for generating the work order?

Written Procedures and Related Records

List records or forms, with their corresponding part numbers, used in this procedure:

List related work instructions and procedures with their corresponding part numbers that employees use as instructions for the activities described above:

CONTRACT REVIEW PROCEDURE TEMPLATE

1. PURPOSE

- *To ensure all inquiries for products are processed efficiently.*
- To ensure the requirements of each contract and amendments to a contract are defined and documented.
- *To maintain contract review records, provide contract authorization, and document the review of contract activities.*
- *To establish procedures for examining and reviewing contract requirements,* including customer-specific requirements, *to ensure that they are adequately defined.*
- *To define the procedures for order entry.*
- *To ensure that requirements for purchase specifications, part numbers, catalog numbers, and terms and conditions are adequately defined.*
- *To provide for customer confirmation, and to ensure company capability in terms of schedule, manufacturing equipment, materials, and personnel.*

2. SCOPE

This procedure applies to products manufactured or sold. These are products in the technical sales literature. Special requests, not addressed in the sales literature, are defined by customer specifications or drawings and reviewed by Marketing and Engineering. Amendments to a contract or purchase order are reviewed and approved using the same process as the original contract or purchase order.

This procedure applies principally to the Sales Department, but may involve every other department of the company.

3. RESPONSIBILITIES

<Job title> generates the quote and sends the approved quote to the customer.

<Job title> reviews and approves quotes, authorizes changes to product pricing, approves procedures and updates, and provides training for new employees.

<Job title> reviews and approves quotes, ensures changes to price lists are made upon approval, forecasts the market, and communicates the customer's needs.

<Department> sets up new accounts and generates sales orders.

<Job title> reviews contract information to track deviations to standard contracts and assigns quote numbers to approved quotes.

<Job title> provides technical advice to customers on the most appropriate product from <Company>'s range for their needs.

<Department> takes orders from customers by telephone, mail, or facsimile (fax), maintains an up-to-date list of customers and addresses, reviews the inventory when an order is placed, and commits stock to orders.

4. PROCEDURE

4.1 General

See Quality System, xxx.

Quotes are generated for the following products:

- *<Product>*
- *<Product>*

Products are priced in company product literature catalogs.

<Job title, Team Name> reviews contracts and/or purchase orders prior to acceptance of a contract or purchase order to ensure that all requirements can be met.

4.2 Marketing Requirements

<Job title> projects the market using <activity>. Customer input is collected <activity>. <Job title> identifies features to put into future products.

4.3 Generating Quotes

<Job title> reviews requests for quotes to determine whether to bid on the quotes. <Job title> tracks a bid through <name of form>. <Job title> maintains the <name of form>. The <name of form> is stored <location>.

<Job title> prepares the quote using <items>. <Job title> estimates material and labor costs and enters the quote information onto <name of form>. <Job title> approves the quote.

<Job title> contacts the customer with an approved quote and sends <title of quotation form> listing <items>.

<Job title> tracks quotations using <item>.

4.4 Order Entry

The types of orders are <order types>.

Upon receipt of a customer order, <job title> records the order onto <name of form>. <Job title> maintains the <name of form> and stores it <location>.

<Job title> assesses if the customer has ordered previously.

<Job title> verifies the following <activities>.

<Job title> establishes credit terms.

<Job title> is responsible for keeping track of customers and maintains customer information <location>.

<Job title> prepares the sales order and is responsible for communicating <written and/or verbal> sales orders to the customer. The customer acknowledges the sales order.

<Job title> maintains the acknowledgments.

Ship dates are provided with the sales order. <Job title> is responsible for estimating the ship date.

<Department> receives the sales order.

For foreign sales, the following documentation is required: <documentation>

If any changes are made to the sales order, <job title> contacts the relevant staff. <Job title> revises the sales order.

4.5 Production Control

Production follows a <type> operating environment. Production levels are determined based on <items>.

<Job title> releases the work order that includes <items>.

Using <software program>, <job title> generates the material requirements, a schedule that lists the raw materials needed, and a capacity plan to assess the available hours per work center and the hours required to manufacture the subassembly or assembly.

5. RELATED DOCUMENTS

<pricing guidelines>
<quote form>
<order entry procedure>
<product literature>

Quality System
Document and Data Control
Handling, Storing, Packaging, Preservation, and Delivery
Quality Records
APQP and Control Plan
Production Planning

<list of work instructions>

Design Control

4.4

What is the job title and name of the person responsible for this procedure?

ISO 9000 Standard:

4.4 Design control

4.4.1 General

The supplier shall establish and maintain documented procedures to control and verify the design of the product in order to ensure that the specified requirements are met.

4.4.2 Design and development planning

The supplier shall prepare plans for each design and development activity. The plans shall describe or reference these activities, and define responsibility for their implementation. The design and development activities shall be assigned to qualified personnel equipped with adequate resources. The plans shall be updated, as the design evolves.

4.4.3 Organizational and technical interfaces

Organizational and technical interfaces between different groups which input into the design process shall be defined and the necessary information documented, transmitted, and regularly reviewed.

4.4.4 Design input

Design-input requirements relating to the product, including applicable statutory and regulatory requirements, shall be identified, documented, and their selection reviewed by the supplier for adequacy. Incomplete, ambiguous, or conflicting requirements shall be resolved with those responsible for imposing these requirements.

Design input shall take into consideration the results of any contract-review activities.

4.4.5 Design output

Design output shall be documented and expressed in terms that can be verified against design-input requirements and validated (see 4.4.8). Design output shall:

a) meet the design-input requirements;
b) contain or make reference to acceptance criteria;

c) identify those characteristics of the design that are crucial to the safe and proper functioning of the product (e.g., operating, storage, handling, maintenance, and disposal requirements).

Design-output documents shall be reviewed before release.

4.4.6 Design review

At appropriate stages of design, formal documented reviews of the design results shall be planned and conducted. Participants at each design review shall include representatives of all functions concerned with the design stage being reviewed, as well as other specialist personnel, as required. Records of such reviews shall be maintained (see 4.16).

4.4.7 Design verification

At appropriate stages of design, design verification shall be performed to ensure that the design-stage output meets the design-stage input requirements. The design-verification measures shall be recorded (see 4.16).

NOTE 10 In addition to conducting design reviews (see 4.4.6), design verification may include activities such as

-performing alternative calculations,
-comparing the new design with a similar proven design, if available,
-undertaking tests and demonstrations, and
-reviewing the design-stage documents before release.

4.4.8 Design validation

Design validation shall be performed to ensure that product conforms to defined user needs and/or requirements.

NOTES

11 Design validation follows successful design verification (see 4.4.7).

12 Validation is normally performed under defined operating conditions.

13 Validation is normally performed on the final product, but may be necessary in earlier stages prior to product completion.

14 Multiple validations may be performed if there are different intended uses.

4.4.9 Design changes

All design changes and modifications shall be identified, documented, reviewed, and approved by authorized personnel before their implementation.

QS 9000 Interpretations and Supplemental Quality System Requirements

The QS 9000 supplements to ISO 9001, 4.4, "Design Control," are:

- Required Skills
- Design Input-Supplemental
- Design Output-Supplemental
- Design Verification-Supplemental
- Design Changes-Supplemental

These additions require that design activities have personnel qualified in specific skills, have CAD/CAE systems capable of two way interface with the customer's CAD/CAE, and are able to control and manage subcontractor activities during design. The supplier must have a comprehensive prototype program that fosters the use of the same subcontractors for production that were used during prototype. The supplier must also obtain customer approval or the customer's waiver of approval for design changes.

Suggested Procedure

QS 9000:

- Advanced Product Quality Planning

Review Team

General

Who (job title/department, i.e., Engineering, Marketing, Manufacturing) translates the customer's needs into technical specifications?

You must hold regular reviews with the necessary organizational and technical groups.

Do you have a formal team (i.e., Design Review Board, Technical Review Team, APQP Team*) to review the specification and ensure that it is predictable, verifiable, and controllable?* ❒*yes* ❒ *no*

If yes, complete the following:

What is the title of the team/board?

Who (job title) comprises the team/board?

When does the team/board meet (i.e., at initial introduction of product specification, at the conclusion of each phase of design development, at final release for production)?

Does the team generate a plan that identifies design activities, and who is responsible? ❒*yes* ❒ *no*

You must prepare design plans that assign responsibility to design activities.

If yes, complete the following:

What is the name of the design plan?

Where is the plan stored?

Is the plan revised as necessary? ❒*yes* ❒ *no*

Are employees who perform design activities qualified? ❒*yes* ❒ *no*

If yes, complete the following:

Do training records identify the employee's qualifications? ❒*yes* ❒ *no*

Who (job title) is responsible for assuring that personnel are qualified to perform design activities?

Do these qualifications include, as appropriate (list by job function, i.e., Manufacturing Engineer, Applications Engineer, Industrial Engineer):

Training	Job Function
Geometric Dimensioning & Tolerancing	
Quality Function Deployment	
Design for Manufacturing/Assembly	
Value Engineering	
Design of Experiments	
Failure Mode Effects Analysis	
Finite Element Analysis	
Solid Modeling	
Simulation Techniques	
Computer Aided Design/Engineering	
Reliability Engineering	

Records

Are minutes of the design meetings maintained as Quality Records? ❒*yes* ❒ *no*

If yes, complete the following:

Who (job title) is responsible for maintaining the records?

Where are the records kept?

How long are the records kept?

Market Review

Who (job title, department, any employee, customer) can initiate a concept for a new product?

Upon what information are features of a new product based (i.e., sales forecasts, industry analysis)?

Is benchmarking used to determine new product features? ❐yes ❐ no

Is the analysis of company-level data used in determining new product features?
❐yes ❐ no

Are customer surveys used to determine new product features? ❐yes ❐ no

Product Definition

You must adhere to Government regulations for country of sale and manufacture.

What are the applicable statutory and regulatory requirements to which your company must adhere (including country of manufacture and countries of sale)?

How are customer requirements communicated to the design team?

Who (job title) assures that all requirements are clear and complete?

Who (job title) develops preliminary drawings?

Are training records for the job above maintained as Quality Records?

❐ *yes* ❐ *no*

Who (job title) approves the preliminary drawings?

If necessary, are models built to demonstrate a new feature? ❐ *yes* ❐ *no*

What documents are generated to support the design (i.e., assembly drawings, bill of materials, test specifications)?

For each of the above, who (job title) is responsible for generating the document? List the document and responsible person:

Are the documents controlled? ❐ *yes* ❐ *no*

Are computer aided product design, engineering, and analysis activities available?
❐yes ❐ no

You must have CAD/CAE capable of interfacing with the customer's system.

If CAD/CAE is subcontracted, how is control over subcontractors maintained(i.e., formal program reviews, on-site representation)?

Is the CAD/CAE capable of two-way interface with the customer's system?
❐yes ❐ no

Who (job title) maintains the CAD/CAE to ensure customer interface capability?

Qualify and Verify

Prototype

Is there a comprehensive prototype program? ❐yes ❐ no

If no, is the requirement waived by the customer? ❐yes ❐ no

Is a prototype of the new design built? ❐*yes* ❐ *no*

If yes, complete the following:

Who (job title) is responsible for testing the new design?

You must identify and document input requirements, resolve any conflicting require-ments, and maintain a comprehensive prototype program.

Who (job title) is responsible for monitoring the test results for conformance to the requirements?

Who (job title) is responsible for tracking the status of the testing to the product development schedule?

Do the performance tests include:

product life testing ❐yes ❐ no
reliability testing ❐yes ❐ no
durability testing ❐yes ❐ no

Performance Testing must be monitored.

Are any of the prototype tests subcontracted? ❐yes ❐ no

If yes, who (job title) is responsible for managing subcontracted technical activities?

What is the title of the record in which the prototype test results are recorded?

Is this record maintained as a Quality Record? ❐*yes* ❐ *no*

If yes, complete the following:

Where is the record stored?

How long is the record kept?

Who (job title, team name) reviews the prototype test results?

If required, are the preliminary design drawings revised? ❐*yes* ❐ *no*

Whenever possible, you must use the same subcontractors during production that were used during prototype.

Are the same subcontractors used for prototype build as production? ❐yes ❐ no

If no, list the rationale for changing subcontractors:

Are the same tooling and processes used for prototype build as production? ❐yes ❐ no

If no, list the rationale for changing tooling or processes:

Who (job title, department) monitors the performance of the subcontractor?

Pilot Run

Are the preliminary drawings released for a pilot run? ❐*yes* ❐ *no*

Who (job title, team name) evaluates the pilot run?

Pilot Test Procedures

You must document the design output to indicate input requirements, acceptance criteria, and regulatory and safety requirements.

Are the test procedures documented? ❐yes ❐ no

If yes, complete the following:

What do the test procedures specify (i.e., the measuring and test equipment used, test schedules, resources)?

What does the evaluation include?

quality function deployment	❐yes	❐ no
value analysis	❐yes	❐ no
design for manufacture/assembly	❐yes	❐ no
design of experiments	❐yes	❐ no
tolerance studies	❐yes	❐ no
response surface methodology	❐yes	❐ no
failure mode and effects analysis	❐yes	❐ no
fault tree analysis	❐yes	❐ no
risk analysis	❐yes	❐ no
regulatory compliance	❐yes	❐ no
safety considerations	❐yes	❐ no
qualification tests	❐yes	❐ no
alternative calculations	❐yes	❐ no
consideration of similar proven designs	❐yes	❐ no
technical test results	❐yes	❐ no
geometric dimensioning & tolerancing	❐yes	❐ no
cost/performance/risk trade-offs	❐yes	❐ no
customer feedback	❐yes	❐ no
other		

What is the title of the record in which pilot test results are recorded?

Is this record maintained as a Quality Record? ❐yes ❐ no

If yes, complete the following:

Who (job title) is responsible for maintaining the record?

Where is the record stored?

How long is the record kept?

Design Review

General

Is there a final review to determine if the product meets the requirements of the product specification? ❒ *yes* ❒ *no*

You must have the design verified using design control methods.

Is there a final review to determine if the product meets the customer's requirements? ❒ *yes* ❒ *no*

If yes, complete the following:

Who (job title) qualifies the pilot run determining if the product meets the requirements of the product specification?

Are subcontracted, suitably qualified facilities used for specialized testing? ❒ *yes* ❒ *no*

Review Criteria

What does the review include?

installation, operation, maintenance, and repair manuals ❒ *yes* ❒ *no*
existence of adequate distribution and customer service organization ❒ *yes* ❒ *no*
training of field personnel ❒ *yes* ❒ *no*
availability of spare parts ❒ *yes* ❒ *no*
field trial ❒ *yes* ❒ *no*
certification of the satisfactory completion of qualification test ❒ *yes* ❒ *no*
physical inspection of early production units, packaging, and labeling ❒ *yes* ❒ *no*
identification of proper handling and storage ❒ *yes* ❒ *no*
maintenance requirements ❒ *yes* ❒ *no*
disposal requirements ❒ *yes* ❒ *no*
evidence of process capability ❒ *yes* ❒ *no*
design stage documents ❒ *yes* ❒ *no*
other

Review Record

What is the title of the form that releases the design to production?

Who (job title) approves the form?

Is this form maintained as a Quality Record? ❐*yes* ❐ *no*

If yes, complete the following:

Who (job title) is responsible for maintaining the record?

Where is the record stored?

How long is the record kept?

Requalification

You must maintain procedures for modifying the design.

Is there periodic reevaluation of product to ensure that the design is still valid with respect to all specified requirements? ❐*yes* ❐ *no*

If yes, complete the following:

Does the review include:

review of customer needs ❐*yes* ❐ *no*
technical specification in light of field experiences ❐*yes* ❐ *no*
field performance surveys ❐*yes* ❐ *no*
new technologies and techniques ❐*yes* ❐ *no*
process modifications ❐*yes* ❐ *no*
production or field experience indicating the need for design change ❐*yes* ❐ *no*
other

What records of the review are maintained?

You must review and approve changes to a design prior to making the change.

Are these records stored as Quality Records? ❒ *yes* ❒ *no*

If yes, complete the following:

Who (job title) is responsible for maintaining the records?

How long are the records kept?

Where are the records stored?

Revision Control

Are all design changes given the same review as the initial design? ❒ *yes* ❒ *no*

Who maintains the documentation?

Who reviews changes and modifications to the design of the product?

Who (job title) approves changes and modifications to the design of the product?

For proprietary designs, what analysis is conducted to assess the effect of the change of form, fit, function, performance, and durability?

Who (job title) conducts these assessments?

Are records maintained of these assessments? ❒ *yes* ❒ *no*

Who (job title) maintains the records?

Written Procedures and Related Records

List records or forms, with their corresponding part numbers, used in this procedure:

List related work instructions and procedures, with their corresponding part numbers, that employees use as instructions for activities described above:

DESIGN CONTROL PROCEDURE TEMPLATE

1. PURPOSE

- *To control the design and development of new products.*
- *To define the checks and balances applied to the design and development activity.*
- *To control and verify the design of the products, assign design function responsibilities to qualified personnel, define technical interfaces, ensure that the product meets the specified requirements, verify that the design output meets the design input requirements and ensure that design changes are properly reviewed before being implemented into production.*
- *To expose the product design to persons with viewpoints and opinions other than those of product design and development engineers.*
- *To reevaluate distributed products.*
- *To maximize protection against oversight that might adversely affect product quality, safety, and efficiency.*

2. SCOPE

This procedure applies to the development of all new products from the initial design to the release for manufacturing. This procedure also applies to design changes to existing product.

3. RESPONSIBILITIES

<Job title> approves the product specification for major new products, product revisions, and accessories.

<Department> evaluates performance, durability, safety, reliability, and maintainability of the design under expected storage and operational conditions; verifies that all design features are as intended and that all authorized design changes have been accomplished and recorded; and, validates computer systems and software.

<Department> provides initial product requirements, performance target values, and market utility assessments.

<Department> ensures the availability of products and services required for the manufacturing and testing of the product. <Job title> monitors the tests for schedule and performance requirements conformance.

<Department, Job title> advises and reviews necessary testing, inspection requirements, and manages subcontractor testing.

<Department> ensures that sufficient personnel are available and suitable for the production of the product, and that an appropriate training program is instituted.

<Department> ensures that the facilities, instrumentation, machinery, equipment, and services are available for the efficient production of the product.

<Department> tracks revision levels and changes to preliminary drawings. <Job title> approves revisions to preliminary drawings.

<Job title> manages CAD/CAE activities and assures that the CAD/CAE system is capable of interfacing with the customer's CAD/CAE system.

<Job title> monitors subcontractor's performance during prototype.

<Job title> coordinates:

- *the design and production of sketches, drawings, and layouts*
- *the building and testing of models, prototypes, and pilots, including the acquisition of materials*
- *the collection and recording of test data*
- *any modifications to the product specification, if necessary*
- the review of requirements for completeness and clarity

At all stages of development, <team> representing <departments> may determine whether to modify the product specification or change the design of the product based on test results.

<Job title> coordinates the appropriate customer approvals for design changes.

4. PROCEDURE

4.1 Review Team

See Contract Review, xxx.

See Advanced Product Quality Planning and Control Plan, xxx.

<Job title> translates the customer's needs into technical specifications.

<Team>, composed of <job titles>, reviews the specification and ensures it is producible, verifiable, and controllable.

<Team> generates a design plan identifying activities and assigning responsibility to the activities. Those assigned to perform design activities meet company qualifications. <Records of qualification> are stored as Quality Records at <location>.

<Team> meets <activities>.

<Records> of <team> meetings are maintained as Quality Records. <Job title> maintains the records for <length of time> in <location>.

4.2 Market Review

See Advanced Product Quality Planning and Control Plan, xxx.

See Management Responsibility, xxx.

<Job title> can initiate a concept for a new product. Features of a new product are based upon <information>.

4.3 Product Definition

See Document and Data Control, xxx.

See Training, xxx.

<Job title> assures that qualified personnel perform design activities. Records of employees' qualifications are maintained by <job title>.

<Job title> develops preliminary drawings. <Job title> approves preliminary drawings.

Customer requirements are incorporated into the design. <Department> communicates the requirements to the design team through <activity>. <Job title> assures that these requirements are clear and complete.

<Job title> ensures that the product design adheres to the applicable statutory and regulatory requirements of the country of manufacture and countries of sale.

Computer Aided Design and Engineering (CAD/CAE) are utilized when appropriate by qualified personnel. <Job title> assures that subcontracted CAD/CAE activities are controlled by

- <Activity>
- <Activity>

The CAD/CAE system at <Company> is maintained by <job title> to ensure two-way interface with the customer's system.

<Job title> generates <document> to support the design. The documents are controlled and maintained by <department>.

4.4 Qualify and Verify

See Quality Records, xxx.

See *Purchasing*, xxx.

Prototype programs include comprehensive design verification activities. *<Job title> is responsible for testing/analyzing the design and records test results on <name of form>*. When appropriate, product life, reliability, and durability testing is part of the test program. The <name of form> is maintained by <department> for <length of time>.

<Job title> tracks and monitors the testing to ensure schedule and performance requirements are met.

<Job title> reviews the test results and, if required, the preliminary design drawings are revised.

When possible, the same subcontractors that provided material and services for prototype are used for pilot runs and production. <Job title/Department> tracks the performance of the subcontractor .

The preliminary drawings are released for a pilot run. <Job title> evaluates the pilot run.

The test procedures specify the following <criteria> and include: <types of analysis>.

<Job title> records the test results onto <name of form>. The <name of form> is maintained by <department> for <length of time>.

4.5 Final Review

A final review to determine if the product meets the requirements of the product specification is held by <team>.

<Job title> qualifies the pilot run determining if the product meets the requirements of the product specification.

Subcontracted, suitably qualified facilities are used for specialized testing.

The review includes <criteria>.

The <name of form> releases the design to production. <Job title> approves the <name of form>. The <name of form> is maintained by <department> for <length of time>.

4.6 Requalification

Periodically, the product is reevaluated to ensure that the design is still valid with respect to all specified requirements. The review includes: <criteria>.

Records of the requalification are maintained on the <name of form>. <Job title> approves the <name of form>. The <name of form> is maintained by <department> for <length of time>.

4.7 Revision Control

Design changes are given the same review as the initial design. <Job title> reviews changes and modifications to the design of the product. <Job title> revises the documentation.

<Job title> coordinates applicable changes with the customer and obtains the necessary customer approval or waiver of approval. The <name of form> is used to document this action.

5. RELATED DOCUMENTS

<records of meetings>
<design plan>
<design reviews>
<form for qualification>
<test procedure>

<training records>

<customer change approval form>

Contract Review
Document and Data Control
Corrective and Preventive Action
Quality Records
Training
Advanced Product Quality Planning and Control Plans
Production Part Approval Process

<list of job instructions>

Notes:

Document and Data Control

4.5

What is the job title and name of the person responsible for this procedure?

ISO 9000 Standard:

4.5 Document and data control

4.5.1 General

The supplier shall establish and maintain documented procedures to control all documents and data that relate to the requirements of this American National Standard including, to the extent applicable, documents of external origin such as standards and customer drawings.

NOTE 15 Documents and data can be in the form of any type of media, such as hard copy or electronic media.

4.5.2 Document and data approval and issue

The documents and data shall be reviewed and approved for adequacy by authorized personnel prior to issue. A master list or equivalent document-control procedure identifying the current revision status of documents shall be established and be readily available to preclude the use of invalid and/or obsolete documents.

This control shall ensure that:

a) the pertinent issues of appropriate documents are available at all locations where operations essential to the effective functioning of the quality system are performed;
b) invalid and/or obsolete documents are promptly removed from all points of issue or use, or otherwise assured against unintended use;
c) any obsolete documents retained for legal and/or knowledge-preservation purposes are suitably identified.

4.5.3 Document and data changes

Changes to documents and data shall be reviewed and approved by the same functions/organizations that performed the original review and approval, unless specifically designated otherwise. The designated functions/organizations shall have access to pertinent background information upon which to base their review and approval.

Where practicable, the nature of the change shall be identified in the document or the appropriate attachments.

QS 9000 Interpretations and Supplemental Quality System Requirements

The QS 9000 supplements to ISO 9001, 4.5, "Document Control," are:

- Reference Documents
- Document Identification for Special Characteristics
- Engineering Specifications

These additions require the supplier to have the most current edition of customer reference documents available at the manufacturing location, to mark documents with the appropriate notation for special characteristics, to review customer engineering specification/standards and changes in a timely manner, and to keep a record of the date that the change is implemented into production.

Suggested Procedures

QS 9000:

- Review of Customer Engineering Standards/Specifications and Changes
- Special Characteristics

Types of Documents

Which documents are under your Document Control system? Use the following table as a guideline.

You must maintain procedures to control all documents related to ISO/QS 9000.

Type of Document	Description
QS 9000 Quality Manual	Describes the quality policy
Quality and operational procedures	Instructions on who does what, when, and where
Work or job instructions	Instructions on how to build and package the product
Routings	*A list of operations showing the flow of product through manufacture*
Rework procedures	*Instructions for reworking already-built product*
Specifications (i.e., product, material, packaging)	*Drawings, blueprints*
Bill of materials	*List of components and items packaged with product or needed to build the product*
Test specifications	*Specification of parameters tested*
Product literature	*Parts catalogs, specification sheets*
Technical publications	*User manuals, reference manuals*
Reference documentation	*Compliance documents*
Service manuals	*Instructions for maintenance and repair*
Audit procedures	*Instructions for performing internal quality audits*
Inspection Instructions	*Instructions on how to inspect product*
Statistical techniques	*Instructions for collecting, analyzing, and reviewing statistical data*
Handling procedure	*Instructions for handling product and components*

Corrective action procedures	*Instructions for identifying root causes of problems*
Customer-supplied product procedures	*Instructions for verification, storage, and maintenance of customer-supplied product*
Purchasing procedures	*Instructions for processing purchase orders*
Design procedures	*Instructions for controlling and verifying product design*
Contract review procedures	*Instructions for processing sales orders*
Quality planning procedures	*Instructions for defining the requirements for quality*

You must have the most current edition of customer reference documents.

Are currently released editions of the document under document control available at all appropriate manufacturing locations (i.e., each facility building the product)?

❒yes ❒ no

If yes, list the locations or the central location for customer source documents:

Who (job title) reviews customer drawings, specifications, and standards to identify reference documents?

Who (job title) ensures that these reference documents are obtained and are the latest revision?

You must mark documents with appropriate special characteristics notations.

Are process control guidelines (PFMEAs, control plans) marked with the appropriate symbol to indicate the process step(s) that affect special characteristics?

❒*yes* ❒ *no*

If yes, what are the special characteristic symbols (i.e., customer-specified, company-specified)?

Distribution

Distribution List

You must ensure that the latest revision of documents are available at the required locations.

Is there a Distribution List of holders of controlled copies? ❒*yes* ❒ *no*

If yes, complete the following:

What is the title of the Distribution List?

Where is the Distribution List found (i.e., in a master file, on the document)?

Where is the list stored?

Who (job title) is responsible for maintaining the list?

For what length of time is the Distribution List maintained?

Where is the Distribution List maintained (i.e., on individual document's master index)?

Who (job title) is responsible for removing obsolete documents?

You must remove obsolete documents.

Are obsolete documents retained for legal reasons or for knowledge preservation? ❒*yes* ❒ *no*

If yes, complete the following:

Where are obsolete documents stored?

How are obsolete documents marked?

How are obsolete documents identified (i.e., stamped "obsolete," destroyed)?

Distribution Cover Sheet

Is there a sign-off sheet that accompanies the distribution? ❐ *yes* ❐ *no*

If yes, complete the following:

What is the title of the sign-off sheet?

Where is the completed sign-off sheet stored?

Who (job title) is responsible for maintaining the completed sign-off sheets?

Which controlled documents are distributed with sign-off sheets?

Electronic Distribution

Are controlled documents distributed electronically? ❐ *yes* ❐ *no*

If yes, complete the following:

Is only the latest version of the controlled document available? ❐ *yes* ❐ *no*

How is the recipient notified of the availability of a revised document?

Identification of Controlled Copy

How are documents marked to indicate a controlled copy (i.e., stamped, dated, printed on special paper, marked electronically)?

You must identify controlled copies.

How are documents marked to indicate an uncontrolled copy (i.e., absence of controlled marking, "reference only")?

Who (i.e., any employee) can request an uncontrolled copy?

Are the locations of uncontrolled copies tracked? ❐ *yes* ❐ *no*

Master List

Is there a master list of procedure titles with current revision levels? ❐ *yes* ❐ *no*

You must maintain a master list that indicates the current revision level.

If yes, complete the following:

What is the title of the master list?

Where is the list stored?

Who (job title) is responsible for maintaining the list?

For what length of time is the master list maintained?

Revisions

Change Requests

Who (job title) can initiate a change to a drawing or standard?

What is the title of the form to request a change to a drawing or standard?

Who (job title) reviews the change request to a drawing or standard?

Who (job title, any employee) can initiate a change to a document?

What is the title of the form to request a change to a document?

Who reviews the change request?

Who (job title) reviews requests to change procedures?

Is the change request reviewed to ensure that the change does not affect the level of control required to meet the standard to which the company is registered? ❒ *yes* ❒ *no*

If yes, is the change request confirmed with the appropriate registration authority when required?

Are change requests logged? ❒ *yes* ❒ *no*

If yes, complete the following:

What is the title of the log?

Who (job title) maintains the log?

Change Request Activities

Are there regular meetings to review change requests? ❐ *yes* ❐ *no*

If yes, complete the following:

Are customer engineering, standards/specifications and changes included in these meetings for review? ❐ yes ❐ no

If no, how are customer changes reviewed?

How often are the meetings scheduled?

Who (job title) attends the meetings?

If the change is approved, who (job title) is responsible for updating affected documents?

If necessary due to safety considerations, who (job title) is responsible for contacting the regulatory agency?

What is the title of the form indicating the change (i.e., Engineering Change Notice)?

You must review customer engineering standards/specifications and changes and record the date the change is implemented into production.

Are records maintained that indicate the day that customer changes are implemented into production? ❒yes ❒ no

If yes, what is the title of the record?

Who (job title) is responsible for maintaining those records?

Document Revisions

Answer the following questions for drawings, standards, specifications, and documents:

Who is responsible for making the change (i.e., originator of procedure)?

Who is responsible for approving changes to procedures (i.e., original reviewers)?

Does the revision level of the document increment? ❒*yes* ❒ *no*

What other documents may be affected as a result of the change (i.e., bill of materials, work procedures)?

Who (job title) is responsible for logging the change?

How is the change logged (i.e., database, logbook)?

What is the name of the log?

Change Record

Is there a record of changes? ❒ *yes* ❒ *no*

You must keep a record of revision changes.

If yes, complete the following:

Where is the change record found (i.e., in a master file, on the document)?

Where is this record stored?

Who (job title) is responsible for maintaining the record of changes?

Change Status

Is the status of the change request recorded? ❐ *yes* ❐ *no*

If yes, complete the following:

What is the title of the status record (i.e., ECO Logbook)?

Where is the status record stored?

Who (job title) is responsible for maintaining the change status record?

Who (job title) reviews the status of change requests opened beyond their implementation date?

What is the frequency of this review?

Written Procedures and Related Records

List records or forms used for document control:

List related work instructions and procedures with their corresponding part numbers which employees use as instructions for following the procedures:

DOCUMENT AND DATA CONTROL PROCEDURE TEMPLATE

1. PURPOSE

- *To ensure that only the most recent revisions of documents for manufacturing and assemblies are available to appropriate personnel.*
- *To control that documents requiring changes are revised in a timely fashion and receive the required approvals.*
- *To ensure that the Quality Manual and QS 9000 procedures are of current issue.*
- *To define the method for establishing, approving, changing, maintaining, replacing, and distributing product documents pertaining to agency approvals.*
- To ensure that the date a change is implemented into production is recorded and that all appropriate documentation is updated to reflect the change.
- To ensure that customer drawings, specifications, and reference documents are the currently released editions and are available at the manufacturing location.

2. SCOPE

All internal component specification sheets, material specifications, test procedures, engineering standards, math (CAD) data engineering drawings, blueprints, rework procedures, Printed Circuit Board (PCB) artwork, Manufacturing Engineering instructions (MEIs), regulatory documentation, work instructions, inspection instructions, Original Equipment Manufacturing (OEM) distribution Bill of Materials (BOM), qualification reports, Engineering Change Orders (ECOs), service and installation documents, handling procedures, subcontractor procedures, routings, and operation sheets are controlled by documentation control.

This procedure covers control of the QS 9000 Quality Manual and the quality and operational procedures that are controlled by <department>.

The marketing literature pertaining to all products and services offered by <Company>.

This procedure applies to reference documents. Reference documents cover certification records that include U/L, CSA, TUV, FCC, and FDA. Reference documents also include underlying customer specifications, drawings, or standards.

3. RESPONSIBILITIES

<Job title> oversees the control of all documents, keeping a master list of the location of all documents.

<Job title> is responsible for ensuring all QS 9000 quality and operational procedures *and the Quality Manual are revised and approved as required. <Job title> notifies employees of available revisions.*

<Job title> distributes work orders with newly revised documents according to a distribution list. When a work order is complete, <Job title> destroys the work order and associated documents.

<Job title> reviews documents to ensure that documents with special characteristics are identified with the <appropriate notation>.

<Job title> reviews customer documents to ensure all referenced documents are available at the <manufacturing location or central location>.

<Job title/Team Name> reviews customer engineering standards/specifications and changes. <Job title> records the date the change is implemented into production.

<Job titles> approve newly released documents and revised documents.

Any employee can request a change to a document.

<Job title> plans product literature based on company needs and writes or assigns a writer to develop the text.

<Job title> is responsible for operation and service manuals. <Job title> ensures that the publication is technically correct and usable, and coordinates with the printer to produce the publication.

<Job titles> review drafts of technical manuals and approve final copy to be sent for production.

<Job title> assigns part numbers and revision levels to technical manuals to control changes to the documents.

<Job title> maintains the ISO 9000 Quality Manuals and procedures in a state that reflects the needs of the company.

<Job title> has primary responsibility for control of engineering changes. This includes ECR number assignment, coordination and approval, expediting, reproduction, and distribution and maintenance of the ECR approval and signature list. <Job title> updates the database and bill of materials to reflect the approved change.

<Job title> maintains agency standards and is the conduit for all agency communications. <Job title> stores reference documents.

4. PROCEDURE

4.1 Types of Documents

The following documents are in the Document Control system.

Type of Document	Description

4.2 Distribution

<Job title> ensures that only the latest revisions of documents are available at the required locations.

Customer drawings and specifications are reviewed by <job title> for reference documents. <Job title> ensures that the most current edition of the referenced documents are available at the <manufacturing location>.

<Job title> reviews documents to ensure that special characteristics are identified with the <appropriate notation>.

<Job title> maintains the distribution list. <Job title> disposes of obsolete documents. <Marking> identifies obsolete documents.

Where necessary, obsolete documents may be retained for legal reasons or for knowledge preservation. <Job title> identifies obsolete drawings with <marking>. <Job title> stores retained obsolete documents <location>.

A <name of distribution coversheet> accompanies the distribution and is signed by the recipient. <Job title> maintains completed sign-off sheets.

The following documents are distributed electronically: <documents>. Only the latest version of the controlled document is available. Recipients are notified of revised documents in the following manner: <activity>.

<Job title> maintains the distribution list. The distribution list is available <location>.

All customer changes are sent to <job title>. Customer engineering standards/ specifications and changes are reviewed by <job title> within <frequency> business days of receipt from the customer. <Name of log> tracks the date the change is implemented into production. <Job title> maintains the <name of log>.

4.3 Document Identification

Controlled copies are indicated by <marking>. <Job title> marks controlled copies with <marking>.

Uncontrolled copies are indicated by <marking>. <Job title> can request an uncontrolled copy.

4.4 Master List

<Job title> maintains a Master List of the most recently approved documents with the current revision number. <Job title> stores the Master List.

The Master List is stored <location>.

4.5 Revisions

<Job title> can initiate a change to a document by completing a <name of form>. <Job title> reviews the change request. Change requests are logged in <name of log>.

<Frequency> meetings to review change requests are attended by <job titles>. If the change is approved, <job title> is responsible for updating affected documents. If necessary, due to safety considerations, <job title> contacts the regulatory agency.

<Name of form> indicates the change.

<Job title> is responsible for making the change. <Job title> approves changes to procedures. <Job title> increments the revision level of affected documents. <Job title> logs revised documents in the <name of log>.

4.6 Change Records

A record of changes is found <location>. <Job title> is responsible for maintaining the change record. <Job title> records the status of the change in the <name of log>. <Frequency>, <job title> reviews the status of change requests opened beyond their implementation date.

5. RELATED DOCUMENTATION

<logbook>
<distribution sheet>
<Master List>
<Engineering Change Request/Order (ECR/ECO)>
<ECR logbook>
<Engineering Change Notice (ECN)>
<customer implementation log>
<customer special characteristics designation>

Design Control
Process Control
Inspection and Testing
Handling, Storage, Packaging, Preservation, and Delivery
Quality Records

<list of work instructions>

Notes:

Purchasing

4.6

What is the job title and name of the person responsible for this procedure?

ISO 9000 Standard:

4.6 Purchasing

4.6.1 General

The supplier shall establish and maintain documented procedures to ensure that purchased product (see 3.1) conforms to specified requirements.

4.6.2 Evaluation of subcontractors

The supplier shall:

a) evaluate and select subcontractors on the basis of their ability to meet subcontract requirements including the quality system and any specific quality-assurance requirements;
b) define the type and extent of control exercised by the supplier over subcontractors. This shall be dependent upon the type of product, the impact of subcontracted product on the quality of final product, and, where applicable, on the quality audit reports and/or quality records of the previously demonstrated capability and performance of subcontractors;
c) establish and maintain quality records of acceptable subcontractors (see 4.16).

4.6.3 Purchasing data

Purchasing documents shall contain data clearly describing the product ordered, including where applicable:

a) the type, class, grade, or other precise identification;
b) the title or other positive identification, and applicable issues of specifications, drawings, process requirements, inspection instructions, and other relevant technical data, including requirements for approval or qualification of product, procedures, process equipment, and personnel;
c) the title, number, and issue of the quality-system standard to be applied.

The supplier shall review and approve purchasing documents for adequacy of the specified requirements prior to release.

4.6.4 Verification of purchased product

4.6.4.1 Supplier verification at subcontractor's premises

Where the supplier proposes to verify purchased product at the subcontractor's premises, the supplier shall specify verification arrangements and the method of product release in the purchasing documents.

4.6.4.2 Customer verification of subcontracted product

Where specified in the contract, the supplier's customer or the customer's representative shall be afforded the right to verify at the subcontractor's premises and the supplier's premises that subcontracted product conforms to specified requirements. Such verification shall not be used by the supplier as evidence of effective control of quality by the subcontractor.

Verification by the customer shall not absolve the supplier of the responsibility to provide acceptable product, nor shall it preclude subsequent rejection by the customer.

QS 9000 Interpretations and Supplemental Quality System Requirements

The QS 9000 supplements to ISO 9001, 4.6, "Purchasing," are:

- Approved Materials for Ongoing Production
- Subcontractor Development
- Scheduling Subcontractors
- Restricted Substances

The additions to the ISO 9001 require the supplier to use subcontractors designated by the customer on an approved supplier list. The supplier must also assure that product and materials meet the governmental, environmental, and safety constraints for restricted, toxic, and hazardous materials for the country of manufacture and country of sale. The supplier must develop subcontractor quality systems using Sections 1 and 2 of the QS 9000. Subcontractors must provide 100% on-time delivery to suppliers and the supplier must provide the subcontractor with the proper information to facilitate 100% on-time delivery. The supplier must also track excessive freight charges. The supplier must comply with governmental and safety constraints for restricted and hazardous substances for the product and the processes involved in the manufacture, storage, shipment, and destruction of the product.

Suggested Procedures

- *Subcontractor Assessment*
- *Generating Purchase Orders*
- *Verifying Purchased Product Conformance*

QS 9000:

- Disposal and Handling of Restricted, Toxic, and Hazardous Substances
- Subcontractor Delivery Performance
- Customer-approved Subcontractor List
- Operational Purchasing Procedures

Customer-approved Subcontractor List

Subcontractor List

You must use customer - approved subcontractors from the approved supplier list.

Does the customer(s) provide an approved subcontractor list? ❐yes ❐ no

If yes, complete the following:

Is the customer-approved subcontractor list reviewed prior to acceptance of a purchase order? ❐yes ❐ no

Who (job title, team name) is responsible for reviewing the customer-approved subcontractor list?

Who (job title) maintains the customer-approved subcontractor list?

Where is the list maintained?

If defective material is received from a customer-approved subcontractor, who (job title) notifies the customer?

Is the customer-approved subcontractor notified of the nonconforming product? ❐yes ❐ no

If yes, what form is used to notify the subcontractor?

Who (job title) negotiates with the customer-approved subcontractor?

Who (job title) is responsible for notifying the customer's Materials Engineering Activity of potential additional subcontractors for the material or product?

Is there a form used for notifying the Materials Engineering Activity? ❐yes ❐ no

If yes, what is the name of the form?

Who (job title) is responsible for assuring that formal customer approval is received for additional subcontractors?

Do you approve subcontractors? ❐*yes* ❐ *no*

If yes, complete the following:

Are there different levels of approvals (i.e., full approval, limited approval, or rejected)? ❐*yes* ❐ *no*

If yes, what levels do you define?

What is the title of your subcontractor list(s)?

Where is this list stored?

Who (job title) is responsible for the list?

Is the list a quality record? ❐*yes* ❐ *no*

Does the list specify company names, materials, and services (i.e., calibration) for which the subcontractor has been approved? ❐*yes* ❐ *no*

Does being on the list qualify the subcontractor for ship-to-stock? ❐*yes* ❐ *no*

You must select subcontractors on the basis of their ability to meet your requirements.

You must establish and maintain records of acceptable subcontractors.

Subcontractor Qualification

You must develop subcontractors using QS 9000.

Do you perform subcontractor quality system development using Sections 1 and 2 of QS 9000? ❒yes ❒ no

If yes, what methods do you use for development (i.e., assessments, seminars, self-certification, benchmarking opportunities, on-site assistance)?

Who (job title) assists in subcontractor development?

You must evaluate subcontractors on their abilities to meet your requirements, as well as evaluate their quality systems and quality assurance requirements.

Who (job title) is responsible for qualifying and approving sources?

Does QS 9000 certification or another quality system qualify a subcontractor for the list? ❒yes ❒ no

How is a subcontractor disqualified from the list (i.e., problems with quality, delivery)?

Are subcontractors evaluated for their ability to ensure that their product meets governmental and environmental constraints in the country of manufacture and country of sale? ❒yes ❒ no

Subcontractor Self-evaluation Questionnaires

Are self-evaluations limited to certain types of companies (i.e., non-complex in nature, prominent reputations)? ❒*yes* ❒ *no*

Do you have a subcontractor self-evaluation questionnaire? ❒*yes* ❒ *no*

If yes, complete the following:

What is the title of the subcontractor self-evaluation questionnaire?

Where are these questionnaires stored?

Who (job title) is responsible for maintaining the questionnaire?

Who (job title) evaluates the responses?

How is the subcontractor self-evaluation monitored (i.e., through testing records)?

What happens if the decision to qualify the subcontractor cannot be determined by the questionnaire?

The subcontractor must ensure that its quality system is effective.

Subcontractor History

Is the subcontractor approval based upon previous history? ❒ *yes* ❒ *no*

If yes, complete the following:

What is the title of the form used for rating the subcontractor's history?

Is the subcontractor history collected from any records (i.e., buy card)? ❒ *yes* ❒ *no*

If yes, which records are used?

What criteria are rated (i.e., meeting 100% on-time delivery requirements, *pricing, quality of product, rejection rates)?*

What is the time frame for subcontractor reassessment?

Subcontractor Samples Qualification

Are specific components qualified? ❒ *yes* ❒ *no*

If yes, complete the following:

What information is required from the subcontractor for the evaluation (i.e., subcontractor information, specification sheets, samples, pricing, lead time, usage)?

Who (job title) oversees testing of all proposed components?

If different from above, who (job title) evaluates and approves the component?

How many samples are required?

What is the evaluation based upon (i.e., technical suitability based on drawings)?

What parameters are measured?

What is the title of the form(s) or record(s) that assess the sample?

Where are these records stored?

Who (job title) is responsible for the records?

Upon approval, is a purchase order issued for a pilot run? ❒*yes* ❒ *no*

If yes, complete the following:

Who (job title) monitors the pilot run?

Subcontractor Site Survey

Are subcontractor site surveys performed? ❐*yes* ❐ *no*

If yes, complete the following:

For which type of subcontractors would a survey be performed (i.e., need more information than self-evaluation provides, subcontractors of critical components)?

Does the subcontractor need to show ability to comply with your customers' contracts? ❐*yes* ❐ *no*

Does the subcontractor need to show he/she can meet your needs? ❐*yes* ❐ *no*

Who (job title) performs the site visit?

How often (frequency) are site visits performed?

If audits are performed by the OEM customer, OEM customer-approved second party, or an accredited third-party registrar, are the audit results recognized by the supplier? ❐yes ❐ no

Does the on-site visit assess the facility's ability to meet your requirements? ❐*yes* ❐ *no*

Does the on-site visit review the organization and responsibilities of your subcontractor? ❐*yes* ❐ *no*

Do you use a survey form? ❐*yes* ❐ *no*

If yes, complete the following:

What is the title of the survey?

Where is the survey stored?

Who (job title) is responsible for maintaining the survey?

Who (job title) evaluates the responses?

Who (job title) approves the subcontractor?

Purchase Requisitions

Inventory Purchase

Are inventory items ordered based on schedules? ❒yes ❒ no

If yes, complete the following:

Who (job title or department, i.e., Production Control) provides the schedule?

Who (job title or department, i.e., Purchasing) orders inventory items?

Who (job title or department, i.e., Production Control) generates requisitions for inventory items?

Inventory Schedule

Is there a report that indicates gross requirement and schedule? ❒yes ❒ no

If yes, complete the following:

What is the title of the report (i.e., Material Review Report)?

What information is used to generate the report (i.e., marketing forecasts, current inventory, ship quantities, delivery schedules)?

Quotations

Who (job title) contacts potential subcontractors for quotes?

Are potential subcontractors from your subcontractor list? ❒ *yes* ❒ *no*

What information does the potential subcontractor provide (i.e., price, delivery)?

Is a record completed with the subcontractor bid information? ❒ *yes* ❒ *no*

If yes, complete the following:

What are the titles of the forms completed (i.e., buy card, history card)?

What information is entered onto the above form?

How are bids evaluated (i.e., price, delivery, past history)?

Purchase Order Approvals

Who (job title) completes the purchase order requisition?

Who (job title) approves the requisition?

Do purchases above price levels require additional approval? ❒*yes* ❒ *no*

If yes, complete the following:

What is the price limit?

Who (job title) approves purchase orders above the limit?

Purchasing Data

Purchase Order Data

What is on the purchase order (i.e., definition of the product ordered, and, if applicable, the type, class, style, grade, title issue of specifications, drawings, inspection instructions, qualification of product, procedures, process equipment, personnel, a quality standard, compliance to governmental, environmental, safety, toxic, and hazardous material constraints applicable to the country of manufacture and sale, 100% on-time delivery, access to subcontractor's facilities by the customer)?

The purchase order must clearly describe the product being ordered.

If the purchase is a catalog order, what is referenced on the purchase order (i.e., the part number and any required modifications)?

What documents are attached to the purchase order (i.e., engineering drawings, specifications, parts lists)?

Are drawings checked for their revision level? ❒ *yes* ❒ *no*

Are materials listed on the drawings? ❒ *yes* ❒ *no*

For special orders, does a detailed drawing of tooling and components accompany the purchase order? ❒ yes ❒ no

Is the subcontractor required to indicate the lot code and your company part number, where applicable? ❒ yes ❒ no

Who (job title) is responsible for assuring that purchased materials conform to current governmental and safety constraints, and meet environmental, electrical, and electro-magnetic considerations for the applicable country of manufacture and sale?

Are records maintained of subcontractor compliance to current governmental, safety, environmental, electrical, and electromagnetic constraints applicable to the country of manufacture and sale? ❒ yes ❒ no

If yes, complete the following:

What is the name of the record?

Who (job title) maintains the record?

Do subcontractors maintain records of conformance to current governmental, safety, environmental, electrical, and electromagnetic constraints applicable to the country of manufacture and sale? ❒ yes ❒ no

Changes to Purchase Orders

What happens if an engineering change causes a revision to the specification of an item?

Who (job title) approves changes to purchase orders?

Who (job title) contacts the subcontractor?

Who (job title) enters the change onto the PO?

Verification of Purchase Order

General

How is the order verified (i.e., the subcontractor's quality assurance system, test data, process control records, 100% testing by the subcontractor, lot acceptance inspection/testing sampling by the subcontractor)?

Where applicable, do purchase orders indicate the methods of conformance (i.e., inspection and test data)? ❒*yes* ❒ *no*

Do you verify purchased product at the subcontractor's site? ❒*yes* ❒ *no*

If yes, complete the following:

What verification arrangements are specified?

What is the method of product release?

Is the method of product release identified in purchasing documents? ❒*yes* ❒ *no*

Does your customer need access to your subcontractor's facility to review product or process? ❒yes ❒ no

If yes, complete the following:

Do you inform subcontractors of this requirement? ❒yes ❒ no

If yes, how is the subcontractor notified (i.e., purchase order, phone call, letter)?

Receiving

Does the recipient evaluate the shipment for transit damage? ❒yes ❒ no

Does the recipient verify the purchase order for:

completeness	❒yes	❒ no
compliance to specified requirements	❒yes	❒ no
documentation	❒yes	❒ no
required certifications	❒yes	❒ no
quantity	❒yes	❒ no
nomenclature	❒yes	❒ no
part number	❒yes	❒ no
other		

Receiving Record

Does the recipient complete any form? ❐ *yes* ❐ *no*

If yes, complete the following:

What is the title of the record?

Where is the record stored?

How long is the record stored?

Who (job title) is responsible for maintaining the record?

Receiving Status

Is an accept/reject receiving status identified on the received part? ❐ *yes* ❐ *no*

If yes, complete the following:

In what form is the status reported (i.e., tag, report)?

What is indicated with the status (i.e., MRR number, report number)?

Scheduling Subcontractors

You must require 100% on-time shipment from subcontractors.

Do you require 100% on-time delivery performance from subcontractors? ❐ yes ❐ no

If yes, are planning documents (i.e., production schedules, inventory replacement levels) provided to subcontractors? ❐ yes ❐ no

What is the name of the planning document?

How frequently is the planning document updated for the subcontractor?

Who (job title) is responsible for updating the planning document?

Who (job title) is responsible for providing the document to the subcontractor?

What is the commitment form that the subcontractor completes to assure the supplier that 100% on-time delivery will be met?

Is there a system to monitor delivery performance of subcontractors? ❒yes ❒ no

You must monitor the delivery performance of your subcontractors.

If yes, complete the following:

What is the name of the system?

Does the system monitor excessive freight charges?

Does the system monitor premium freight charges?

Who (job title) maintains the system?

What is the name of the record that tracks delivery performance?

How frequently is the record updated?

What is the recourse if a subcontractor does not meet 100% on-time delivery?

Verification of Purchased Product

If your customer needs access to the subcontractor's facilities, it must be stated in the purchasing documentation.

Does the customer need access to a subcontractor's facility to review product or process?
❒yes ❒ no

If yes, complete the following:

Is the subcontractor informed of this requirement? ❒yes ❒ no

If yes, how is the subcontractor notified (i.e., purchase order, phone call, letter)?

Who (job title) informs the subcontractor of this requirement?

Does the customer need access to a subcontractor's facility to inspect and accept product?
❒yes ❒ no

If yes, complete the following:

Is the subcontractor informed of this requirement? ❒yes ❒ no

If yes, how is the subcontractor notified (i.e., purchase order, phone call, letter)?

Who (job title) informs the subcontractor of this requirement?

Are records maintained of the customer's visit? ❒yes ❒ no

If yes, complete the following:

What is the name of the record?

Who (job title) maintains the record?

How long is the record maintained?

Disputes

General

Who (job title) dispositions defective or nonconforming parts?

Who (job title) negotiates with the subcontractor if nonconforming materials are received?

Are subcontractors immediately informed of the nonconformance and asked to furnish a corrective action proposal? ❒*yes* ❒ *no*

If a nonconforming shipment is received, is there a form that is completed? ❒*yes* ❒ *no*

If yes, complete the following:

Who (job title) completes the form?

What is the title of the form?

Review of Defective or Nonconforming Parts

Are defective shipments reviewed? ❒*yes* ❒ *no*

If yes, complete the following:

Who (job title, department, or group, i.e., Material Review Board) reviews the parts?

Is the subcontractor reviewed? ❒*yes* ❒ *no*

If yes, what is the title of the record that indicates the review?

Does the review affect the subcontractor's approval? ❒*yes* ❒ *no*

If yes, explain:

Written Procedures and Related Records

List records or forms maintained when processing purchase orders and assessing subcontractor:

List related work instructions and procedures with their corresponding part numbers that employees use as instructions for processing purchase orders and assessing subcontractors:

PURCHASING PROCEDURE TEMPLATE

1. PURPOSE

- *To ensure that purchased product conforms to the specified requirements.*
- *To ensure that parts ordered from the manufacturer's catalog meet engineering requirements.*
- *To ensure that purchase orders clearly define the product ordered.*
- *To select subcontractors on their ability to meet the company's requirements.*
- *To keep records of acceptable subcontractors and ensure that quality system controls are effective for subcontractors.*
- *To incorporate purchased product into the supplies.*
- To ensure that product is purchased from the customer-approved subcontractor list.
- To ensure the development of subcontractors' quality systems to Sections I and II of QS 9000.
- To ensure the quality of subcontracted parts, materials, and services.
- To ensure 100% on-time delivery from subcontractors.
- To establish a system to monitor subcontractor delivery performance.
- To ensure that purchased parts meet government and safety constraints on restricted, toxic, and hazardous substances for the country of manufacture and country of sale.

2. SCOPE

This procedure applies to all materials and parts ordered for incorporation into the products, consumed during the production process, or used to operate the business.

3. RESPONSIBILITIES

<Job title> reviews purchase orders, based on the amount of the order.

<Job title> rates subcontractors who have a previous history.

<Job title> surveys potential suppliers' facilities.

<Job title> determines whether to negotiate new purchase orders or amend existing orders, based on the manufacturing schedule.

<Job title> ensures all items ordered are adequately defined through specifications, drawings, and parts lists to guarantee that requirements are met.

<Job title> writes and approves all purchasing procedures and is responsible for their distribution.

Purchasing:

- *Selects a supplier capable of meeting the requirements with respect to quality, timely delivery, and cost.*
- *Orders materials to meets the production schedule.*
- *Orders expense items, upon the supervisor's approval.*
- *Maintains codes in the software system and categorizes ordered items by cost and type.*
- *Participates in <name of team> meetings.*
- *Reorders items received in inferior quality, if so dispositioned.*

<Job title> performs incoming inspection, as required.

<Job title> prepares the <name of report> for inventory purchases.

<Job title> verifies the purchase order for quantity and delivery requirements.

<Job title> monitors subcontractor's on-time delivery performance.

<Job title> verifies that product and material meet all safety, governmental, and environmental concerns for the country of manufacture and country of sale.

4. PROCEDURE

4.1 Customer-approved Subcontractor List

The following lists the types of approvals: <approval types> for the <name of approved subcontractor list>. <Job title> maintains the list in <location>.

The <name of approved subcontractor list> specifies company names, and materials and services for which the subcontractor has been approved. Being on this list qualifies the subcontractor for ship-to-stock.

4.2 Customer-approved Subcontractor List

<Job titles/Department>purchase product from the customer-approved supplier list. <Job title> maintains the list and the list is stored in <location>. The list is reviewed <frequency> by <job title> to assure that it is current and accurate.

Additional subcontractors may be added to the list upon successful completion of:

- <Activity>
- <Activity>

<Job title> submits the proposed subcontractors to the customer's engineering activity for approval. Once approved, the subcontractor is added to the list.

4.3 Governmental and Environmental Constraints

<Job title> reviews material lists, specifications, and other relevant data to ensure that material used in the manufacture and packaging of the product meets current governmental and safety constraints on restricted, toxic, and hazardous materials, as well as electric and electromagnetic considerations. This review includes the country of manufacture, as well as the country of sale.

If material does not meet the criteria either for the country of manufacture or the country of sale, <job title>:

- <Activity>
- <Activity>

<Job title> ensures that restricted, toxic, and hazardous substance safety constraints are followed for purchased product and manufacturing.

4.4 Subcontractor Qualification

Subcontractors are qualified for the list in the following manner:

- QS 9000-certified, automatic qualification
- *Self-evaluation questionnaire*
- *Subcontractor history*
- *Sample qualification*
- *Site survey*

- Customer audit
- Customer-approved second-party audit

<Job title> qualifies and approves subcontractors. A subcontractor who is QS 9000-certified qualifies for the <approved subcontractor list>. A subcontractor is disqualified from the list by <activity>.

A subcontractor completes the <name of self-evaluation questionnaire>. <Job title> evaluates the responses. The subcontractor is monitored through <activity>.

<Job title> rates the subcontractor's previous history on the <name of form>. The history is compiled from <records> and rates the following criteria: <criteria>. <Frequency>, subcontractors are reassessed.

Specific components are evaluated for approval. <Job title> oversees testing of all proposed components. <Job title> evaluates and approves the components. A minimum on <number> samples are required for evaluation. The evaluation is based on <criteria> and measures the following parameters: <parameters>. <Job title> records the evaluation onto <name of form>. The <name of form> is stored <location>.

Upon approval, a purchase order is issued for a pilot run. <Job title> monitors the pilot run.

<Job title> surveys the subcontractor's site to determine the subcontractor's ability to comply with contracts and the subcontractor's ability to meet needs. The subcontractor's organization is reviewed. The results of the survey are recorded onto <name of survey>. The completed survey is stored <location>. Customer, customer second-party approved, and registrar audits are accepted in lieu of supplier surveys.

<Job title> evaluates the subcontractor's responses. <Job title> approves the subcontractor for the <name of subcontractor vendor list>.

4.5 Purchase Requisitions

See *Contract Review, xxx* and *Process Control*, xxx.

<Job title> generates the schedule for which to purchase inventory items. Based on the schedule, <job title> generates the requisitions for the items.

The <name of report> indicates the gross requirement and schedule. Information from <activity> is used to generate the report.

4.6 Quotations

<Job title> contacts potential suppliers from the approved subcontractor list for a quote and requests <information> from the supplier.

<Job title> completes <record> with the supplier's bid information. <Job title> evaluates the bids based on <information>.

4.7 Purchase Order Approvals

<Job title> completes the purchase order requisition. <Job title>approves the requisition. Purchase above <price> requires additional approval of <job title>.

If an engineering change causes a revision to the specification of an item <job title> contacts the subcontractor. <Job title> enters the change onto the purchase order.

The order is verified by <activity>. Where applicable, purchase orders indicate the following methods of conformance: <items>.

4.8 Purchasing Data

The purchase order indicates <information>.

If the purchase is a catalog order, <information> is referenced on the purchase order.

The purchase order defines <information>.

<Documents> are attached to the purchase order.

<Job title> checks drawings attached to the purchase order for the correct revision level. Materials are listed on the drawings.

For special orders, a detailed drawing of tooling and components accompanies the purchase order.

The purchase order specifies that the supplier is required to indicate the lot code and company part number, where applicable.

When the purchase order requires customer inspection and/or acceptance of product at the subcontractor's location, <form> is completed by the subcontractor and returned to <job title/department>.

4.9 Verification of Purchase Order

See *Inspection and Testing*, xxx and *Handling, Storage, Packaging, Preservation, and Delivery*, xxx.

Upon receipt of an order, <job title> verifies the quantity and completes the <name of form>.

Upon receipt of the parts, <job title> evaluates the shipment for transit damage and verifies the order for: <list>.

The receiver completes <name of form>. <Job title> stores the <name of form> in <location> for <length of time>.

<Job title> identifies the accept/reject receiving status on the received part by <marking>. The <marking> shows the <identification number>.

For items not approved for stock, <job title> inspects the items for <parameter> and completes the <name of record>.

If the inventory item is approved for stock, it is received directly to the stockroom.

<Purchased parts> are verified at the subcontractor's site. <Job title> inspects the parts according to <test plan>. <Job title> identifies the verification status on the inspected part by <marking> and completes the <name of form>.

4.10 Scheduling Subcontractors

See Packaging, Handling, Shipping, and Storage, xxx.

See Corrective and Preventive Action, xxx.

<Job title> provides subcontractors <list documentation> for production and shipment planning. The subcontractors are monitored by <job title> for 100% on-time performance.

<Job title> maintains a log of delivery performance. If a subcontractor does not maintain 100% on-time performance, the subcontractor must <list activities> to remain as a supplier.

The <name of log> is located <location> and includes excessive freight charges, including premium freight charges. <Job title> issues a <corrective action request> when excessive freight charges occur. The subcontractor must provide a <corrective action form> response.

4.11 Disputes

See *Control of Nonconforming Product*, xxx and *Corrective and Preventive Action*, xxx.

<Job title> dispositions defective or nonconforming parts.

<Job title> negotiates with the subcontractor of nonconforming materials received. Subcontractors are immediately informed of the nonconformance and asked to furnish a corrective action proposal.

If a nonconforming shipment is received, <job title> completes the <name of form>.

<Job title> reviews defective shipments.

<Job title> reviews the subcontractor and completes the <name of form>. The subcontractor's rating for the approved subcontractor list is affected.

5. RELATED DOCUMENTS

<purchase order>
<product specification>
<inspection requirements>
<drawings>
<request for quotation>
<approved subcontractor list>
<subcontractor delivery performance log>
<customer acceptance of product at subcontractor facility report>

Contract Review
Document and Data Control
Process Control
Inspection and Testing
Control of Nonconforming Product
Corrective and Preventive Action
Handling, Storage, Packaging, Preservation, and Delivery

<list of work instructions>

Control of Customer-Supplied Product

4.7

What is the job title and name of the person responsible for this procedure?

ISO 9000 Standard:

4.7 Control of Customer-Supplied Product

The supplier shall establish and maintain documented procedures for the control of verification, storage, and maintenance of customer-supplied product provided for incorporation into the supplies or for related activities. Any such product that is lost, damaged, or is otherwise unsuitable for use shall be recorded and reported to the customer (see 4.16).

Verification by the supplier does not absolve the customer of the responsibility to provide acceptable product.

QS 9000 Interpretations and Supplemental Quality System Requirements

The QS 9000 supplements ISO 9001, 4.7, "Control of Customer Supplied Product," by adding customer-owned tooling and returnable packaging as part of customer-supplied product.

General

Does your company use customer-supplied product (including tooling and returnable packaging)?

❒*yes* ❒ *no*

If no, you do not need to write a procedure for this element.

If yes, what product(s) does your company have that are supplied by the purchaser?

Receiving

Answer this section for receiving information. Part III addresses incoming inspection.

Who (job title) is responsible for receiving the material?

What is verified:

purchase order for part number, description, and quantity	❐*yes*	❐ *no*
condition of packaging	❐*yes*	❐ *no*

What is the title of the record that verifies receiving?

What is included in this record (i.e., date, inspections)?

Who (job title) approves the record?

What type of identification is placed on the customer-supplied product indicating the receiving status?

If a problem is found, who (job title) informs the purchaser of the nonconformance?

How is the customer notified (i.e., letter, phone call)?

Who (job title) determines the corrective action?

What records of the corrective action are maintained?

Who (job title) is responsible for ensuring that the corrective action is implemented?

Incoming Inspection

For parts that require inspection, is the material inspected against quality and engineering criteria? ❐*yes* ❐ *no*

If yes, complete the following:

What level of inspection do you provide (i.e., AQL #, 100%)?

What is the title of the record that verifies inspection?

What is included in this record (i.e., date, inspections)?

Who (job title) approves the record?

What type of identification is placed on the customer-supplied product indicating the status of incoming inspection?

If a problem is found, who (job title) informs the purchaser of the nonconformance?

Who (job title) determines the corrective action?

What records of the corrective action are maintained?

Who (job title) is responsible for ensuring that the corrective action is implemented?

In-process Production Control

Do you have a written procedure indicating how to identify the customer-supplied product? ❒*yes* ❒ *no*

If yes, complete the following:

What is the title of the procedure?

Who (job title) is responsible for generating the procedure?

Is the procedure under document control? ❒*yes* ❒ *no*

Do you maintain inspection records for each step in the process requiring inspection? ❒*yes* ❒ *no*

If yes, complete the following:

What is the title of the inspection record?

Who (job title) performs the inspection or test?

What type of identification is placed on the customer-supplied product indicating the status of in-process inspection?

If a problem is found, who (job title) determines the corrective action?

What records of the corrective action are maintained?

Who (job title) is responsible for ensuring that the corrective action is implemented?

Storage Control

Identification

What type of stores control is used to identify the location of customer-supplied product (i.e., logbook, database)?

Who (job title) is responsible for maintaining and identifying the locations of stored products?

What type of identification is applied to the customer-supplied product?

Environment

How is access to the storage area controlled (i.e., locked, restricted)?

Do you have procedures for controlling product that requires special environmental conditions (i.e., control of humidity, temperature, pest, electromagnetic fields, security, fire)? ❒ *yes* ❒ *no*

If yes, complete the following:

What are the titles of the procedures?

Who (job title) is responsible for generating the procedure?

Is the procedure under document control?

Shelf Life

Is the shelf life of the product monitored? ❐yes ❐ *no*

If yes, complete the following:

How is the product dated (i.e., labels, database, bar-coded)?

Who (job title) is responsible for ensuring that shelf life has not expired?

Storage Inspection

Is stored product periodically inspected to detect deterioration? ❐yes ❐ *no*

If yes, complete the following:

Who (job title) is responsible for inspecting the product?

What is the title of the inspection record?

Who (job title) approves the inspection record?

Are the records stored as quality records?

Security

How is security of customer-supplied product maintained (i.e., product must be signed in and out, authorized)?

What records do you have indicating the locations of products (i.e., log, database)?

Who (job title) reviews the records to ensure that all product that is signed out is returned?

Control of Damaged or Lost Product

In the event of damage to the product, do you follow a written procedure? ❐*yes* ❐ *no*

If yes, complete the following:

What is the title of the procedure?

Who (job title) is responsible for generating the procedure?

Is the procedure under document control? ❐*yes* ❐ *no*

What is the title of the form used for recording the damage or loss?

What is included on the form (i.e., part number, description of damage or loss, date of occurrence)?

Is corrective action implemented to prevent recurrence and/or to fix the problem? ❐*yes* ❐ *no*

If yes, complete the following:

Who (job title) informs the purchaser of the nonconformance?

Who (job title) determines the corrective action?

What records of the corrective action are maintained?

Who (job title) is responsible for ensuring that the corrective action is implemented?

Do you have adequate insurance in place in accordance with the customer's contract to cover the loss or damage? ❐ *yes* ❐ *no*

Written Procedures and Related Records

List records or forms, with their corresponding part numbers, used in this procedure:

List related work instructions and procedures, with their corresponding part numbers, that employees use as instructions for activities described above:

CONTROL OF CUSTOMER-SUPPLIED PRODUCT PROCEDURE TEMPLATE

1. PURPOSE

- *To ensure customer-supplied product is protected against loss or deterioration.*
- *To maintain records of customer-supplied product.*

2. SCOPE

This procedure applies to product, tooling, and returnable shipping material supplied by the customer to support the manufacture, assembly, or shipment of product. Customer-supplied product includes product supplied by the customer for incorporation into the customer's product or to support the manufacture, assembly, or shipment of the customer's product (i.e., tooling and returnable containers).

3. RESPONSIBILITIES

<Job title> inspects customer-supplied product upon receipt of the item.

<Job title> inspects customer-supplied product during inprocess production.

<Job title> ensures customer-supplied product is safe from deterioration, damage, or loss.

4. PROCEDURE

4.1 General

The following items are supplied by the customer:

- *<Item>*
- *<Item>*

4.2 Receiving

<Job title> is responsible for receiving customer-supplied material, verifies the purchase order for part number, description, and quantity, and verifying the condition of the packaging.

<Job title> records the receiving information onto the <name of form> and enters the date and <inspection data>. <Job title> approves the <name of form>.

The receiving status of the customer-supplied product is identified by <marking>.

If a problem is found during receiving, <job title> informs the customer of the nonconformance. <Job title> determines the corrective action and records the corrective action onto <name of form>.

<Job title> is responsible for ensuring the corrective action is implemented.

4.3 Incoming Inspection

For parts that require inspection, the material is inspected against quality and engineering criteria to <inspection level>.

The inspection is recorded onto <name of form> and includes <information>. <Job title> approves the <name of form>.

The inspection status of the customer-supplied product is identified by <marking>.

If a problem is found during the inspection, <job title>informs the customer of the nonconformance. <Job title> determines the corrective action and records the corrective action onto <name of form>.

<Job title> is responsible for ensuring that the corrective action is implemented.

4.4. In-process Production Control

<Name of procedure> describes how the customer-supplied product should be identified. <Job title> is responsible for generating the procedure. The procedure is in the document control system.

Inspection records for each step in the process requiring inspection are maintained. <Job title> performs the inspections and records the results onto <name of form>.

The inspection status of the customer-supplied product is identified by <marking>.

If a problem is found during the inspection, <job title>informs the customer of the nonconformance. <Job title> determines the corrective action and records the corrective action onto <name of form>.

<Job title> is responsible for ensuring that the corrective action is implemented.

4.5 Storage Control

The storage location of customer-supplied product is recorded on <name of form>. <Job title> is responsible for maintaining and identifying the location of stored product. The customer-supplied product is identified by <marking>.

Access to the storage area is controlled by <activity>. Procedures for controlling product that requires special environmental conditions are generated by <job title>.

The shelf life of the product is monitored using <marking>. <Job title> is responsible for ensuring that shelf life has not expired.

<Frequency>, <job title> inspects stored product to detect deterioration. The inspection is recorded onto <name of form>. <Job title> approves the inspection form. The inspection form is stored <location>.

4.6 Security

The security of customer-supplied product is maintained through <activity>. <Name of record> indicates the location of product. <Job title> reviews the records to ensure that all product that is signed out is returned.

4.7 Control of Damaged or Lost Product

In the event of damage to the product, the <name of procedure> is followed. <Job title> is responsible for generating the procedure. The <name of procedure> is in the document control system.

Damage or loss is recorded onto <name of form> and includes <information>.

Corrective action is implemented to prevent recurrence and/or to fix the problem.

<Job title> informs the customer of the nonconformance. <Job title> determines the corrective action and records the corrective action onto <name of form>.

<Job title> is responsible for ensuring that the corrective action is implemented.

Insurance is in place in accordance with the customer's contract to cover loss or damage.

5. RELATED DOCUMENTS

<receiving record>
<inspection record>
<procedure for damaged product>

Contract Review
Purchasing
Inspection and Testing
Corrective and Preventive Action
Handling, Storage, Packaging, Preservation, and Delivery

<list of work instructions>

Notes:

Product Identification and Traceability

4.8

What is the job title and name of the person responsible for this procedure?

ISO 9000 Standard:

4.8 Product identification and traceability

Where appropriate, the supplier shall establish and maintain documented procedures for identifying the product by suitable means from receipt and during all stages of production, delivery, and installation.

Where and to the extent that traceability is a specified requirement, the supplier shall establish and maintain documented procedures for unique identification of individual product or batches. This identification shall be recorded (see 4.16).

QS 9000 Interpretations and Supplemental Quality System Requirements

The QS 9000 supplements ISO 9001, 4.8, "Product Identification and Traceablity," explain that the "where appropriate" in the above paragraph applies to situations where the identity of the product is not "inherently obvious."

General

Are you required by contract to provide records of traceability? ❐ *yes* ❐ *no*

Do you impose your own requirements for traceablity? ❐ *yes* ❐ *no*

You must document your procedures for maintaining product identification.

Do you have written procedures to identify product from receipt through production, delivery, and installation? ❐ yes ❐ no

If yes, complete the following:

Who (job title) writes the procedures?

Are identification methods described in work instructions? ❐ *yes* ❐ *no*

Which product(s) are traceable through serialization?

Are parts traceable to subcontractors? ❐ *yes* ❐ *no*

If yes, what is the method (i.e., purchase order, receiving inspection records)?

What parts are traceable to subcontractors?

Component Parts

Collecting Identification

Do you have records to show and record traceability? ❐ *yes* ❐ *no*

If yes, complete the following:

What is the name of the record (History Card, Traveler Card)?

What does the record indicate (i.e., subcontractor, date of purchase, part number, serial number)?

Who (job title) records the information onto the card?

Do you use a barcode scanner to record the serial number of a part? ❐*yes* ❐ *no*

Marking the Parts

Are identification numbers recorded onto product parts?

If yes, complete the following:

Who (job title, operator) records the numbers?

Where are the numbers recorded (i.e., onto unit housing, component parts)?

Labeling the Parts

Do you apply identification labels to component parts? ❐*yes* ❐ *no*

If yes, complete the following:

What does the label indicate (serial number, part number, revision level)?

Who (job title) is responsible for affixing the label (i.e., operator, subcontractor)?

If the revision level is applied, is it traceable to drawings and associated ECOs? ❐*yes* ❐ *no*

Traceability through Bill of Materials

If applicable, complete the following:

Can the Bill of Materials be cross-referenced to the drawing for tracing the component to the subcontractor? ❐ *yes* ❐ *no*

Can component parts be tracked using the Bill of Materials and monthly shipments? ❐ *yes* ❐ *no*

Is this information maintained on a database? ❐ *yes* ❐ *no*

Traceability through Warehouse Management System

If applicable, complete the following:

Do you use an automated Warehouse Management System? ❐ *yes* ❐ *no*

If yes, complete the following:

Are items received from the subcontractor, entered into the system through the purchase order number, and assigned an inventory location? ❐ *yes* ❐ *no*

When items are issued from inventory, does the Warehouse Management System track the location of the pick? ❐ *yes* ❐ *no*

Serialization

General

What do you serialize (i.e., assemblies, subassemblies)?

List the parts that are serialized:

Who (job title) is responsible for assigning serial numbers?

Logging Serial Numbers

Are serial numbers logged? ❒*yes* ❒ *no*

If yes, complete the following:

Where are the serial numbers logged (i.e., logbook, software)?

What is the name of the log (i.e., Serial Logbook, MRP software)?

Who (job title) is responsible for entering the serial number (i.e., department, Production Control)?

Applying Serial Numbers

Who (job title) assigns the serial number (i.e., operator)?

How are the serial numbers obtained (i.e., stickers given to operator, serial number listing supplied with shop packet)?

Recording Serial Numbers

How are serial numbers recorded from the unit (i.e., serial history card that travels with the unit, barcode scanner at points in assembly)?

Traceability to Customer

Is the finished product traceable to the customer? ❐ *yes* ❐ *no*

If yes, complete the following:

Is the serial number assigned to a sales order? ❐ *yes* ❐ *no*

What is the method of traceability (i.e., database records, quality records)?

Does the packing list include the serial number of the unit? ❐ *yes* ❐ *no*

Is a serial number applied to the packaging? ❐ *yes* ❐ *no*

Who (job title) verifies that the serial number on the packing list matches the serial number on the box?

Product Identification

How is product identified throughout all stages of production (i.e., tag system, traveler that accompanies product)? If product identification varies from department to department, list the method used in each department:

Written Procedures and Related Records

List records or forms, with their corresponding part numbers, used in this procedure:

List related work instructions and procedures, with their corresponding part numbers, that employees use as instructions for activities described above:

PRODUCT IDENTIFICATION AND TRACEABILITY PROCEDURE TEMPLATE

1. PURPOSE

- *To identify a product from the receipt of material through product installation.*
- *To maintain historical records of the item.*
- *To be able to trace assemblies and selected subassemblies through serialization.*
- *To define the methods which must be used to provide product identification and traceability integrity by means of labeling, barcoding, and serialization.*
- *To provide procedures for identifying and tracking major components, modules, test results, and final product during all stages of production and delivery.*

2. SCOPE

This procedure applies to all parts manufactured and sold.

This procedure ensures that adequate controls are designed into the manufacturing process to provide identification as required throughout the process.

3. RESPONSIBILITIES

<Job title> assigns serial numbers to assemblies and maintains the completed record.

<Job title> keeps track of serial numbers by maintaining the database.

<Job title> ensures that proper labels are present at each operation requiring a label.

<Job title> ensures that labels are properly stored and identified to prevent mix-ups.

4. PROCEDURE

4.1 Written Procedures

Product identification from receipt through production, delivery, and installation is described in <name of operational procedure(s) or work instruction>.

<Job title> applies <identification marking> according to the <work instruction>.

4.2 Component Parts

Where required by contract, component parts are traceable to the manufacturer.

<Name of record> displays a record of traceability. The <name of record> indicates <items>. <Job title> enters the information onto the <name of record> through <activity>. <Job title> applies identification numbers onto the product parts.

<Job title> applies labels to component parts. The label indicates <items>. The revision level is identified on the label and is traceable to drawings and associated ECOs through <activity>.

Component parts can be tracked using the <documents> and <frequency> shipments. This information is maintained <location>.

The <name of system> tracks items received from the vendor by associating the purchase order number with an inventory location and pick.

4.3 Serialization

The following items are serialized: <items>. <Job title> is responsible for assigning serial numbers.

<Job title> enters serial numbers into the <name of logbook>.

<Job title> gives the serial numbers to <job title> who applies the serial number to the part. <Job title> <method> the serial numbers from the unit onto <name of record>.

4.4 Traceability to Customer

Finished product is traceable to the customer through serial numbers associated with <name of customer record>. The packing list and outside packaging include the serial number of the unit. <Job title> verifies that the serial number on the packing list matches the serial number on the box.

5. RELATED DOCUMENTS

<bill of materials>
<serial number logbook>
<history card>
<database>
<packing list>
<sales order>

Contract Review
Document and Data Control

Process Control
Inspection and Test Status
Handling, Storage, Packaging, Preservation, and Delivery
Quality Records

<list of work instructions>

Notes:

Process Control

4.9

What is the job title and name of the person responsible for this procedure?

ISO 9000 Standard:

4.9 Process control

The supplier shall identify and plan the production, installation, and servicing processes which directly affect quality and shall ensure that these processes are carried out under controlled conditions. Controlled conditions shall include the following:

a) *documented procedures defining the manner of production, installation, and servicing, where the absence of such procedures could adversely affect quality;*
b) *use of suitable production, installation, and servicing equipment, and a suitable working environment;*
c) *compliance with reference standards/codes, quality plans, and/or documented procedures;*
d) *monitoring and control of suitable process parameters and product characteristics;*
e) *the approval of processes and equipment, as appropriate;*
f) *criteria for workmanship, which shall be stipulated in the clearest practical manner (e.g., written standards, representative samples, or illustrations);*
g) *suitable maintenance of equipment to ensure continuing process capability.*

Where the results of processes cannot be fully verified by subsequent inspection and testing of the product and where, for example, processing deficiencies may become apparent only after the product is in use, the processes shall be carried out by qualified operators and/or shall require continuous monitoring and control of process parameters to ensure that the specified requirements are met.

The requirements for any qualification of process operations, including associated equipment and personnel (see 4.18), shall be specified.

NOTE 16 Such processes requiring prequalification of their process capability are frequently referred to as special processes.

Records shall be maintained for qualified processes, equipment, and personnel, as appropriate (see 4.16).

QS 9000 Interpretations and Supplemental Quality System Requirements

The QS 9000 supplements to ISO 9001, 4.9, "Process Control," are:

- Government Safety and Environmental Regulations
- Designation of Special Characteristics
- Preventative Maintenance
- Process Monitoring and Operator Instructions
- Preliminary Process Capability Requirements
- On-going Process Performance Requirements
- Modified Preliminary of On-going Capability Requirements
- Verification of Job Set-ups
- Process Changes
- Appearance Items

These additions require the supplier to maintain control of processes through all stages of development. The supplier must perform preliminary and on-going process capability studies that meet the customer's requirements. Special characteristics must be identified and controlled and process change effective dates must be recorded. The supplier must handle and dispose of hazardous waste according to governmental and environmental requirements. Job set-ups must be verified and documentation must be available to the personnel performing the set-up. Appearance items must be evaluated by qualified personnel. The integrity of the masters must be maintained.

Suggested Procedures:

- *Process Control*
- *Temporary Deviations*
- *Production Hold*
- *New Product Introduction*
- *Equipment Maintenance*

QS 9000:

- Control Plans
- Process Capability Studies
- Process Monitoring and Work Instructions
- Appearance Items
- Hazardous Material

General

Who (job title) develops the written manufacturing procedures?

Who (job title) determines the required equipment?

Who (job title) determines the required operator skills?

Written Procedures

Do the procedures indicate workmanship criteria? ❐*yes* ❐ *no*

If yes, complete the following:

How is the workmanship indicated (i.e., written standards, models, samples, photographs, videos)?

You must stipulate workmanship criteria.

Are the examples under document/data control? ❐*yes* ❐ *no*

If yes, complete the following:

Who (job title) is responsible for maintaining the examples of workmanship?

You must document procedures for work that affects quality.

List the types of documentation used for process control, who (job title) generates the document, and indicate (yes/no) if the document is under Document Control. Use this list as a reference.

Documentation	Responsible Person	Document Control (yes/no)
Engineering drawings		
Functional specs		
Instruction sheets		
Manufacturing engineering instructions		
Routings		
Bill of materials		
Test specifications		
Work instructions		
Engineering change notices		
NC machine instructions		
Setup sheets		
Servicing records		
Other		

Who (job title) is responsible for ensuring that only the current revision is distributed to the shop floor?

Training

Types of Training

What type of training is provided for operators (i.e., on-the-job, hired with expertise)?

Who (job title) certifies that an operator is qualified to assemble parts?

Are skill levels classified based on experience and training? ❏*yes* ❏ *no*

If yes, describe the skill levels.

Do you provide special training? ❏*yes* ❏ *no*

If yes, complete the following:

What processes require special training (i.e., repair, soldering, reliability testing)?

Describe the special training.

Who (job title) certifies the recipient of special training?

Records

Are training records kept? ❐yes ❐ no

If yes, complete the following:

What is the title of the training record?

What do the records identify (i.e., part numbers, operators, work centers)?

Do training records identify primary and secondary operators qualified to perform the work? ❐yes ❐ no

Where are the records stored?

Who (job title) is responsible for maintaining the records?

Who (job title) signs and approves the training records?

Control of Environment

Complete the following, indicating the areas in which you have controls in place and the types of control.

Area of Control	Implementation
ESD	*ESD mats, wrist straps, heel straps*
Temperature / humidity	
Waste disposal	
Protective apparel	
Other	

Control of Equipment

You must control your equipment.

Does your equipment comply with product safety requirements? ❐*yes* ❐ *no*

If yes, complete the following:

Which certificates of compliance do machines carry (i.e., UL, CSA, FCC, TUV)?

Does your company comply with governmental and environmental regulations, including the handling, recycling, elimination and disposal of hazardous material?

❐yes ❐ no

If yes, complete the following:

Which certificates or letters of compliance are maintained?

Who (job title) maintains the records of compliance?

Where are the compliance records stored?

How long are the compliance records kept?

Do you have any equipment that requires evaluation prior to its being used?

❐*yes* ❐ *no*

Preventative Maintenance

Is preventative maintenance performed on equipment? ❐*yes* ❐ *no*

You must have a preventative maintenance system.

Is there an effective total planned preventative maintenance system in place?
❐yes ❐ no

If yes, complete the following:

Who (job title) is responsible for preparing and maintaining the written procedure describing the preventative maintenance system?

Who (job title) is responsible for scheduling maintenance activities?

What records for the equipment are verified (i.e., calibration)?

Who (job title) is responsible for ensuring that the equipment is verified?

Where are the records maintained?

Who (job title) is responsible for maintaining the machinery?

What records of preventative maintenance are kept?

How often are the machines maintained?

What predictive maintenance methods are used (i.e., tool wear, review of manufacturer's recommendations, correlation of SPC data to maintenance action)?

You must use predictive maintenance methods.

Are replacement parts available for key manufacturing equipment? ❒yes ❒ no

Who (job title) is responsible for identifying key process equipment?

Who (job title, outside source) ensures the repair of equipment, when necessary?

Process Revision Control

Permanent Changes

You must control your processes.

Who (job title, any employee) can initiate a change to a part or process?

What is the title of the form to complete to initiate a change?

Who (job title) reviews the form?

How is the responsibility to resolve the problem implemented (i.e., corrective action meetings, supervisor delegates)?

What is the time frame assigned for resolution?

Who (job title) is responsible for resolving the problem?

Is the resolution reviewed? ❐ *yes* ❐ *no*

If yes, complete the following:

Who (job title) approves the resolution?

What records are kept indicating the resolution (i.e., ECO, ECN, *Procedure Change Notice)?*

Where is the record kept?

Who (job title) is responsible for ensuring changes to documentation?

Temporary Deviations

Who (job title, any employee) can initiate a temporary change?

What is a temporary deviation used for (i.e., test parts or processes, part replacement, specific time period)?

What is the title of the form to complete to initiate a temporary deviation?

Who (job title) approves the temporary deviation?

Can temporary deviations be converted to permanent changes? ❐*yes* ❐ *no*

If yes, complete the following:

Who (job title) approves the permanent change?

What records are maintained indicating the permanent change (i.e., ECO, ECN, Procedure Change Notice)?

Are temporary deviations extended when necessary? ❐*yes* ❐ *no*

Who (job title) approves the extension?

Is a status report of deviations maintained? ❐*yes* ❐ *no*

If yes, complete the following:

Who (job title) prepares the status report?

What is the title of the status report?

Who (job title/committee) reviews the status of deviations?

Are expired deviations reviewed to determine closure? ❐*yes* ❐ *no*

Production Hold

Do you have a procedure for initiating a stop to production if necessary? ❒*yes* ❒ *no*

If yes, complete the following:

Who (job title/Any employee) has the authority to initiate a hold to production or shipment because of a problem?

Is a form completed for the production hold? ❒*yes* ❒ *no*

If yes, complete the following:

What is the title of the form?

What information is entered onto the form?

Who (job title) approves the form?

Who (job title) maintains the completed form?

Is the completed form entered into a log? ❒*yes* ❒ *no*

If yes, complete the following:

What is the name of the log?

Who (job title) is responsible for maintaining the log?

What is the time frame for issuing a stop to production?

Who (job title) is responsible for issuing a stop to production?

Who (job title) is responsible for resolving the problem?

What records are kept as a result of the corrective action (i.e., ECO, Deviation Notice)?

Who (job title) is responsible for resuming production?

Who (job title) measures the progress of the resolution?

Verify Process

General

How are steps in the process verified (i.e., inspections, quality checks, tests)?

You must monitor and control your processes.

What is the quantity verified (100%, sampling)?

Where applicable, are processes verified for compliance with:

reference standards or codes	❒*yes*	❒ *no*
quality plans	❒*yes*	❒ *no*
written procedures	❒*yes*	❒ *no*

Who(job title/department, i.e., operator, Quality, group leader) is responsible for verifying the work?

What is the frequency of inspection (i.e., completion of job, end of shift)?

What is inspected?

form	❒*yes*	❒ *no*
fit	❒*yes*	❒ *no*
mechanics	❒*yes*	❒ *no*
diagnostic tests	❒*yes*	❒ *no*
mechanical	❒*yes*	❒ *no*
electrical	❒*yes*	❒ *no*
functional	❒*yes*	❒ *no*
aesthetics	❒*yes*	❒ *no*
other		

What records are maintained to indicate verification (i.e., signed routings, barcode scan, checklists)?

Do records require approval signature? ❒*yes* ❒ *no*

If yes, complete the following:

Who (job title) approves the inspection?

What record is kept and for how long?

Who (job title) is responsible for maintaining the record?

Verification of Process

Are data collected on problem processes? ❒*yes* ❒ *no*

If yes, complete the following:

What types of data are collected (i.e., first-pass yield, fall-off)?

Who reviews the data?

Are processes audited? ❒*yes* ❒ *no*

If yes, who (job title) performs the audit?

Is there an audit plan?

If yes, is there a procedure documenting the audit plan? ❐ *yes* ❐ *no*

Who (job title) reviews audit findings?

Do corrective actions follow your procedure for corrective action? ❐ *yes* ❐ *no*

If no, describe the procedure that is implemented.

Special Processes

What special processes are used in your facility (i.e., soldering, welding, reliability testing)?

How are processes verified (visual inspection, SPC)?

Are written procedures available for these processes? ❐ *yes* ❐ *no*

If yes, complete the following:
Who (job title) is responsible for generating the procedures?

Are the procedures under Document Control? ❐ *yes* ❐ *no*

What training for special processes is provided?

Who (job title) provides the training?

What is the title of the records of special training?

Who (job title) signs the training record?

Where are the records kept?

How long are the records stored?

Who (job title) is responsible for maintaining the records?

Process Monitoring and Operator Instructions

You must have documented work instructions accessible at the work-station.

Are there documented process monitoring and operating instructions for each process?
❐yes ❐ no

If yes, complete the following:

List the manufacturing, shipping, receiving, and assembly processes in your facility in the first column. In the second column, annotate (yes/no) whether there are documented process monitoring and operating instructions for each process. In the third column, list the type of instruction (i.e., process sheet, shop traveler, standard work instruction sheet, control plan). In the fourth column, list (yes/no) whether the instructions are available at the workstation.

Process	Instructions	Type	Available at Workstations(s)
Receiving			
Material handling			
Warehouse			
Staging			
Line ____			
Line ____			
Line ____			
Line ____			
Line ____			
Line ____			
Line ____			
Line ____			
Shipping			

Are process flow charts available for all operations? ❑yes ❑ no

If yes, complete the following chart for all processes listed from above:

Do the process monitoring (work instructions) include the following information?

Process Name: ________________

operation name and number keyed to process flow chart	❑yes	❑ no
part name and number	❑yes	❑ no
current engineering level/date	❑yes	❑ no
required tools, gages, and other equipment	❑yes	❑ no
material identification and disposition instructions	❑yes	❑ no
customer- and supplier-designated special characteristics	❑yes	❑ no
spc requirements	❑yes	❑ no
relevant engineering and manufacturing standards	❑yes	❑ no
inspection and test instruction	❑yes	❑ no
correction action instructions	❑yes	❑ no
revision date and approvals	❑yes	❑ no
visual aids	❑yes	❑ no
tool change intervals and set-up instructions	❑yes	❑ no

Process Name: ________________

operation name and number keyed to process flow chart	❑yes	❑ no
part name and number	❑yes	❑ no
current engineering level/date	❑yes	❑ no
required tools, gages, and other equipment	❑yes	❑ no
material identification and disposition instructions	❑yes	❑ no
customer- and supplier-designated special characteristics	❑yes	❑ no
spc requirements	❑yes	❑ no
relevant engineering and manufacturing standards	❑yes	❑ no
inspection and test instruction	❑yes	❑ no
correction action instructions	❑yes	❑ no
revision date and approvals	❑yes	❑ no
visual aids	❑yes	❑ no
tool change intervals and set-up instructions	❑yes	❑ no

Process Name: ________________

operation name and number keyed to process flow chart	❐yes	❐ no
part name and number	❐yes	❐ no
current engineering level/date	❐yes	❐ no
required tools, gages, and other equipment	❐yes	❐ no
material identification and disposition instructions	❐yes	❐ no
customer- and supplier-designated special characteristics	❐yes	❐ no
spc requirements	❐yes	❐ no
relevant engineering and manufacturing standards	❐yes	❐ no
inspection and test instruction	❐yes	❐ no
correction action instructions	❐yes	❐ no
revision date and approvals	❐yes	❐ no
visual aids	❐yes	❐ no
tool change intervals and set-up instructions	❐yes	❐ no

Process Name: ________________

operation name and number keyed to process flow chart	❐yes	❐ no
part name and number	❐yes	❐ no
current engineering level/date	❐yes	❐ no
required tools, gages, and other equipment	❐yes	❐ no
material identification and disposition instructions	❐yes	❐ no
customer- and supplier-designated special characteristics	❐yes	❐ no
spc requirements	❐yes	❐ no
relevant engineering and manufacturing standards	❐yes	❐ no
inspection and test instruction	❐yes	❐ no
correction action instructions	❐yes	❐ no
revision date and approvals	❐yes	❐ no
visual aids	❐yes	❐ no
tool change intervals and set-up instructions	❐yes	❐ no

Preliminary Process Capability Requirements

Are preliminary process capability studies conducted for supplier-designated special characteristics?

❑yes ❑ no

Do you comply with customer requirements for designating, marking, and controlling special characteristics?

Are preliminary process capability studies conducted for customer-designated special characteristics?

❑yes ❑ no

You must perform preliminary process capability studies for special characteristics.

If yes, complete the following:

What is the Ppk target value for these preliminary studies?

How do you assure that Ppk values meet the customer requirements?

What action is taken if the Ppk values do not meet customer and/or supplier requirements (contact customer, implement mistake-proofing activities)?

On-going Process Performance Requirements

You must perform on-going process capability studies.

Are on-going process capability studies performed? ❐yes ❐ no

If yes, complete the following:

What process capability values are required for the following conditions:

- Stable process and normally distributed data ________
- Chronically unstable processes with output meeting specification and a predictable pattern ________

How is on-going process capability determined for non-normal data (i.e., parts per million)?

Are control plans used for all special characteristics? ❐yes ❐ no

If yes, complete the following:

Are there appropriate reaction plans available for characteristics on a control plan that are either noncapable or unstable? ❐yes ❐ no

Are control plans modified when process capability requirements are changed?

Are significant process events (i.e., tool change, machine repair, set-up changes) noted on control charts?

How is continuous improvement implemented for all processes regardless of demonstrated capability?

Continuous improvement is required for all processes.

Verification of Job Set-ups

Are documented instructions available for the set-up of all processes? ❐yes ❐ no

You must verify that the job set-up will produce conforming product.

If yes, complete the following:

Who (job title) is responsible for preparing the set-up instructions?

Who (job title) is responsible for verifying the set-up instructions?

How are changes to set-up procedures recorded and approved?

Is set-up verification performed in accordance with customer requirements? ❐yes ❐ no

Are last off comparisons used? ❐yes ❐ no

Process Changes

Is there a procedure for submitting process changes for customer approval after production part approval is granted? ❐yes ❐ no

Is there a record of process change effective dates? ❐yes ❐ no

You must maintain a record of process changes and their effective dates.

If yes, complete the following:

Who (job title) maintains these records?

Where are the records maintained?

How long are the records maintained?

How are the process change dates correlated to the appropriate production part drawing level?

Appearance Items

Do you produce any customer-designated appearance items? ❒yes ❒ no

If yes, complete the following:

Is the lighting appropriate in all areas where appearance items are manufactured, tested, or inspected?

Are appropriate masters maintained for color, grain, and texture?

color	❒yes	❒ no
grain	❒yes	❒ no
texture	❒yes	❒ no

Where are these masters maintained?

How is the validity of the masters confirmed?

How is the validity of the equipment that maintains the masters confirmed?

Who (job title) is responsible for maintaining the masters?

Who (job title) is responsible for approving the qualifications and training of the operators performing appearance item verification?

How are personnel qualified to perform the verification of appearance items(i.e., color blindness screening)?

What records are maintained of these qualifications?

Who (job title) is responsible for maintaining these records?

Where are the records kept?

How long are the records stored?

Written Procedures and Related Records

List records or forms, with their corresponding part numbers, used in this procedure:

List related work instructions and procedures, with their corresponding part numbers, that employees use as instructions for activities described above:

PROCESS CONTROL PROCEDURE TEMPLATE

1. PURPOSE

- *To control process through appropriate approvals, verification, and documented instructions and through adequate documentation and training.*
- *To prepare for the production of a part or product by generating a work order for the production and by ensuring delivery of parts, the work order (with necessary documentation), and any machine instructions.*
- *To schedule production of assemblies and subassemblies to meet anticipated customer needs, as determined by sales and marketing.*
- *To initiate a change in a process that is not meeting 100% quality by providing a mechanism to deviate temporarily from accepted standards for materials, processes, components or products and a means by which any employee can stop a process or shipment when 100% quality is not being achieved.*

2. SCOPE

This procedure applies to:

- *The manufacture and assembly of piece-part, subassemblies, and final assemblies from the issuance of a work order through packaging.*
- *The rework of nonconforming product.*
- *Temporary changes to the manufacturing process which may be required when parts are no longer available or the design of a part requires change due to an improper fit or malfunction.*

3. RESPONSIBILITIES

<Job title> creates the routings, and bill of materials, and approves ECOs. <Job title> assembles ECO kits and updates associated documents. <Job title> revises drawings and specifications when a process requires change due to improper fit or the malfunction of an assembly or subassembly.

<Job title> maintains the drawings, prints, and specifications and attaches these documents to the work order, ensuring the current revision levels of the attachments.

<Job title> provides on-the-job training.

<Job title> supervises the assembly and ensures controls are in place.

<Job title> plans production and ensures appropriate inventory levels.

<Job title> issues work orders outlining the sequence of manufacturing steps required for the completion of each manufacturing job.

<Job title> orders raw materials listed on the Materials Requirement Plan.

<Job title> ensures that the quality levels required by contract specification are in place and that compliance to documentation is met.

<Job title> inspects operations, assemblies, and subassemblies that are manufactured or processed to ensure that the product meets the product specification.

<Job title> controls a temporary change to the process.

<Job title> prepares process monitoring and work instructions.

<Job title> prepares control plans.

<Job title> conducts process capability studies.

<Job title> maintains appearance masters.

<Job title> evaluates appearance items.

4. PROCEDURE

4.1 Written Procedures

See *Document and Data Control, xxx.*

<Job title> develops the written manufacturing procedures, determines the required equipment, and determines the required operator skills.

<Job title> prepares the process monitoring and operator instructions and assures that they are accessible at the workstation.

The procedures indicate workmanship criteria using <items> which are under document control. <Job title> is responsible for maintaining the examples of workmanship. <Job title> is responsible for ensuring only the current revision is distributed to the shop floor.

The following is a list of documentation used for process control and the person responsible for the document.

Documentation	Responsible Person
Engineering drawings	
Functional specs	
Process monitoring instruction sheets	
Manufacturing engineering instructions	
Routings	
Bill of Materials	
Test specifications	
Work instructions	
Engineering Change Notices	
NC machine instructions	
Set-up sheets	
Control plans	
Preliminary process capability studies	
On-going process capability studies	
Operator instructions	

4.2 Training

See *Training, xxx.*

<Job title> provides training. Trainers are certified through <activity>. The <name of training record> identifies <items>. <Job title> approves and signs the training record. The training record is stored <location>.

Operators are trained <activity>. <Job title> certifies that an operator is qualified to assemble parts. The following lists the classification of skill levels:

- *<Level>*
- *<Level>*
- *<Level>*

Special training is provided for <items>. The training involves <description>. <Job title> certifies that special training has been received.

4.3 Process Capability

<Job title> performs preliminary process capability (Ppk) studies for each process with a special characteristic and compares the Ppk to the customer's requirement. If a Ppk is not specified by the customer, a Ppk $\geq$ 1.67 is the target value.

On-going process capability (Cpk) studies are conducted by <job title> and monitored to assure that the process is stable and capable according to the customer's requirements. If no customer requirements are specified, the following apply:

- Stable, capable process and normally distributed data, then a Cpk≥ 1.33.
- Chronically unstable process, product within specification and a predictable pattern, then a Ppk≥ 1.67.

<Job title> requests customer review of capability requirements when a process demonstrates continuing stability and capability or a lack of stability and capability. The control plan is annotated after the customer's concurrence on actions that need to be taken.

4.4 Special Characteristics

<Job title, Team Name> identifies special characteristics on the control plan according to the customer-supplied designation system.

4.5 Control of Environment

The following indicates the areas in which controls are in place and the types of controls that apply.

Area of Control	Implementation
ESD	*ESD mats, wrist straps, heel straps*
temperature / humidity	
waste disposal	
protective apparel	
other	

Procedures for the handling, recycling, and elimination or disposal of hazardous materials are prepared by <job title>.

<Job title> assures that <Company> is in compliance with applicable governmental and environmental regulations.

4.6 Control of Equipment

See *Inspection, Measuring, and Test Equipment, xxx.*

The process machinery carries the following compliance <list>. <Job title> maintains the records of compliance. The records of compliance are stored <location> for <length of time>.

The following equipment requires evaluation prior to being used: <equipment>. Records verifying the evaluation are maintained by <job title> and are stored <location> for <length of time>.

<Job title> is responsible for preventive maintenance on the equipment. Records of preventive maintenance are maintained by <job title> and are stored <location>.

When necessary, <job title/source> is responsible for ensuring the repair of equipment.

4.7 Process Revision Control

<Job title> can initiate a change to a part or process by completing the <name of form>. <Job title> approves the form.

Responsibility to resolve the problem is assigned at <activity> to be resolved by <time frame>.

<Job title> is responsible for resolving the problem. <Job title> reviews and approves the resolution. <Name of record> indicates the resolution. The <name of record> is stored <location>.

<Job title> is responsible for ensuring changes to affected documentation and maintains a record of process change effective dates.

<Job title> can initiate a temporary change when necessary due to <activity>. <Job title> completes the <name of form> to initiate a temporary deviation. <Job title> approves the <name of form>. The <name of form> is stored in <location> for <length of time>.

4.8 Temporary Deviations

If temporary deviations need to be converted to permanent changes, <job title> approves the permanent changes. Permanent changes are indicated on <documents>.

When necessary, temporary deviations are extended. <Job title> approves the extension.

<Job title> maintains a status report of deviations. The <name of status report> indicates <items>. <Job title> reviews the status of deviations and reviews expired deviations to determine closure.

4.9 Production Hold

<Job title> has the authority to initiate a hold on production or shipment because of a problem.

To initiate a stop on production, <job title> completes the <name of form> and enters the following information: <data>.

<Job title> approves the form. <Job title> stores the <name of form> in <location> for <length of time>.

<Job title> logs the <name of form> into the <name of logbook>.

<Job title> issues the stop on production within <time frame>. <Job title> is responsible for resolving the problem. The <name of record> indicates the result of the corrective action.

<Job title> is responsible for resuming production.

<Job title> measures the progress of the resolution through <activity>.

4.10 Verify Process

See *Inspection and Test, xxx* and *Corrective and Preventive Action, xxx.*

Job set-up is verified to ensure that parts produced conform to the specifications. Set-up verification is performed according to <activity>.

Steps in the process are verified through <activity>. The verification quantity is <quantity>. <Frequency>, <job title> is responsible for verifying the work and checks <list>.

<Job title> completes the <name of record> to indicate verification. <Job title> approves the inspection record. <Job title> is responsible for storing <name of record> in <location> for <length of time>.

The following data are collected on problem processes: <type of data>. <Job title> reviews the data. <Job title> audits processes according to an audit plan. Corrective actions follow the corrective action process.

Where applicable, <job title> verifies processes to assess compliance to reference standards and codes, to evaluate implementation of quality plans, and to determine adequacy and accuracy of written procedures.

4.11 Special Processes

The following special processes are used in production: <processes>. <Job title> generates written procedures that are controlled by <department>.

Special processes are verified by <activity>.

<Job title> provides training for special processes. The <name of record> identifies the completion of training. <Job title> signs the training record. <Job title> stores training records in <location> for <length of time>.

4.12 Appearance Items

Customer-designated appearance items are evaluated by <job title>.

Areas used for evaluation of appearance items are:

- <Description of area>
- <Description of area>

Masters of appearance items are maintained by <job title> in <location>.

5. RELATED DOCUMENTS

<production plan>
<materials requirement plan>
<master production schedule>
<training record>
<work instructions>
<discrepant materials report>
<inspection data sheet>
<control plans>
<process capability studies>
<appearance masters>
<hazardous waste handling procedure>

Document and Data Control
Inspection and Testing
Inspection, Measuring, and Test Equipment
Inspection and Test Status
Control of Nonconforming Product
Corrective and Preventive Action
Handling, Storage, Packaging, Preservation, and Delivery
Quality Records
Internal Quality Audits
Training
Statistical Techniques

<list work instructions>

Notes:

Inspection and Testing

4.10

What is the job title and name of the person responsible for this procedure?

ISO 9000 Standard:

4.10 Inspection and testing

4.10.1 General

The supplier shall establish and maintain documented procedures for inspection and testing activities in order to verify that the specified requirements for the product are met. The required inspection and testing, and the records to be established, shall be detailed in the quality plan or documented procedures.

4.10.2 Receiving inspection and testing

4.10.2.1 The supplier shall ensure that incoming product is not used or processed (except in the circumstances described in 4.10.2.3) until it has been inspected or otherwise verified as conforming to specified requirements. Verification of the specified requirements shall be in accordance with the quality plan and/or documented procedures.

4.10.2.2 In determining the amount and nature of the receiving inspection, consideration shall be given to the amount of control exercised at the subcontractor's premises and the recorded evidence of conformance provided.

4.10.2.3 Where incoming product is released for urgent production purposes prior to verification, it shall be positively identified and recorded (see 4.16) in order to permit immediate recall and replacement in the event of nonconformity to specified requirements.

4.10.3 In-process inspection and testing

The supplier shall:

a) inspect and test the product as required by the quality plan and/or documented procedures;
b) hold product until the required inspection and tests have been completed or necessary reports have been received and verified, except when product is released under positive-recall procedures (see 4.10.2.3). Release under positive-recall procedures shall not preclude the activities outlined in 4.10.3a.

4.10.4 Final inspection and testing

The supplier shall carry out all final inspection and testing in accordance with the quality plan and/or documented procedures to complete the evidence of conformance of the finished product to the specified requirements.

The quality plan and/or documented procedures for final inspection and testing shall require that all specified inspection and tests, including those specified either on receipt of product or in-process, have been carried out and that the results meet specified requirements.

No product shall be dispatched until all the activities specified in the quality plan and/or documented procedures have been satisfactorily completed and the associated data and documentation are available and authorized.

4.10.5 Inspection and test records

The supplier shall establish and maintain records which provide evidence that the product has been inspected and/or tested. These records shall show clearly whether the product has passed or failed the inspections and/or tests according to defined acceptance criteria. Where the product fails to pass any inspection and/or test, the procedures for control of nonconforming product shall apply (see 4.13).

Records shall identify the inspection authority responsible for release of product (see 4.16).

QS 9000 Interpretations and Supplemental Quality System Requirements

The QS 9000 supplements to ISO 9001, 4.10, "Inspection and Testing," are:

- Acceptance Criteria
- Accredited Laboratories
- Incoming Product Quality
- Prevention-based Process Activities
- Layout Inspection and Functional Testing

These additions require the supplier to use zero defects acceptance criteria for attribute data sampling, to use accredited labs if required by the customer, to use subcontractor performance data as well as product inspection during receiving inspection, to use prevention-based inspection methodology throughout the system, and to conduct layout inspections in accordance with customer requirements.

Suggested Procedures:

- *Receiving Inspection and Testing*
- *In-process Inspection and Testing*
- *Final Inspection and Testing*

QS 9000:

- Customer Layout Inspection
- Control Plans

Acceptance Criteria

Do you use sampling plans? ❐yes ❐ no

If yes, complete the following:

What plan(s) do you use for attribute sampling?

What is the acceptance criteria for attribute sampling (i.e., zero defects)?

If other than zero defects is used as acceptance criteria, do you have customer approval? ❐yes ❐ no

What plans do you use for variable sampling?

Are you required by your customer to use accredited laboratories for any of your testing? ❐yes ❐ no

You need to use accredited laboratories when required by the customer.

If yes, complete the following:

Product	**Lab Used**	**Accreditation Verified (yes or no)**

Test Verification

Are all necessary gauges, tools, meters, instruments, and equipment maintained in the calibration system? ❒*yes* ❒ *no*

If yes, complete the following:

Who is responsible for maintaining the calibration records?

Where are the records stored?

How long are the records kept?

Receiving

See the Purchasing worksheet for component qualification.

Incoming Inspection

Who (job title) performs the incoming inspections?

Who (job title) is responsible for holding parts until the required inspection and tests are complete?

How are parts identified as acceptable/nonacceptable for production?

Is there a particular location where material is held while awaiting the completion of incoming inspections and tests? List separate areas for separate products, if applicable.

Verification Systems

You may consider the amount of control exercised at the subcontractor's premises and recorded evidence of conformance when determining the nature of the receiving inspection. You must use subcontractor-supplied data or evaluation as part of the incoming inspection.

Are parts 100% inspected? ❐ *yes* ❐ *no*

If yes, list the types of items:

Are any parts sample-inspected? ❐ *yes* ❐ *no*

If yes, list the items:

What is the sampling plan used (i.e., Military Standard 105)?

Is the subcontractor required by contract to supply inspection records according to your requirements (i.e., statistical data, second- or third-party assessments of the subcontractor's location, parts evaluation by accredited labs, certifications of conformance)? ❐ yes ❐ no

If yes, complete the following:

What data are used?

Who (job title) reviews the inspection data?

Who (job title) approves the inspection data?

How is the status of the inspection data marked (i.e., stamped, initialed, dated)?

Where is the reviewed inspection data stored?

Do you have a ship-to-stock program? ❐ *yes* ❐ *no*

If applicable, does the subcontractor identify in advance nonconforming material and make special arrangements? ❒ *yes* ❒ *no*

Incoming Inspection Activities

Does the inspector use documentation to verify the received product? ❒ *yes* ❒ *no*

You must verify product in accordance with documented procedures.

If yes, what documentation is used (i.e., drawings, specifications, subcontractor history card, prior inspection reports)?

Are written inspection procedures available for test and inspection? ❒ *yes* ❒ *no*

If yes, what are the title and part numbers of the written inspection procedures?

Records

Is there a form on which the inspector records the inspection results? ❒*yes* ❒ *no*

If yes, complete the following:

What is the name of the report (i.e., History Card, Receiving Report)?

What is recorded onto the report (quantity passed, dimensions)?

Who approves the report?

Where is the report stored?

How long is the report kept?

Is the inspection status indicated on the item? ❒*yes* ❒ *no*

If yes, what marks are applied (i.e., stamp, tag, label)?

Emergency Release

You must identify incoming product released for urgent purposes.

Do you identify product released for urgent production that is not inspected? ❒ *yes* ❒ *no*

If yes, complete the following:

Do you affix a label to the non-inspected component? ❒ *yes* ❒ *no*

Do you record the non-inspected part onto a form? ❒ *yes* ❒ *no*

If yes, what is the name of the form (i.e., Traveler, History Card)?

Where is the form stored?

Incoming Inspection –Defective and Discrepant Parts

How is the inspection status indicated (failure tag, stamp, computer printout)?

Where are defective or discrepant parts held?

Who (job title) dispositions the defective item (department, i.e., Quality, Purchasing, Engineering)?

What form is completed to initiate the disposition (i.e., Discrepant Material Report)?

Is the subcontractor notified of the defective and discrepant material?

In-process Testing

In-process Tests

You must document in-process inspections and tests.

Do you have test procedures? ❒ *yes* ❒ *no*

If yes, complete the following:

Are the test procedures under document control? ❒ *yes* ❒ *no*

Who (job title) is responsible for ensuring that test procedures are current?

What is the test quantity (i.e., 100%, sample)?

How are test results measured (i.e., SPC)?

List and describe the tests performed (i.e., hipot, discharge, functional, diagnostic)

Test	Description	Test Record

How are test results recorded (i.e., barcode scanner, records)?

How is the test status identified on the unit (i.e., traveler, stamp, label, computer printout)?

In-process Test Records

Is there a form on which the operator records the test results? ❒ *yes* ❒ *no*

If yes, complete the following:

What is the name of the report?

What is recorded onto the report?

Who approves the report?

Records must identify the inspection authority for release of product.

Where is the report stored?

How long is the report kept?

You must document conformance to specified requirements.

In-process Inspections

Inspections

Who (job title) is responsible for holding parts until the required inspection and tests are complete?

What do in-process inspections verify?

revision levels	❐ *yes*	❐ *no*
details on specifications	❐ *yes*	❐ *no*
visual inspection	❐ *yes*	❐ *no*
functional inspection	❐ *yes*	❐ *no*
dimensional inspection	❐ *yes*	❐ *no*
other		

Who (job title) is responsible for the verification?

What is the scope of the inspection (i.e., 100%, sample)?

Types of In-process Inspections

List and describe the inspections performed (i.e., first piece pull test, paint inspection, numeric control pieces, sheet metal templates, aesthetics)

Test	Description	Test Record

Verification of Set-up Piece

Do you have a first piece inspection? ❐ *yes* ❐ *no*

If yes, complete the following:

Who (job title) inspects that the set-up piece matches the specifications?

Who (job title) approves the set-up piece?

Is there a signed approval? ❐ *yes* ❐ *no*

If yes, what form is signed (i.e., routing, Inspection Report)?

What happens if the set-up piece is rejected (second set-up piece is made and inspected, failure ticket is attached, item is placed in holding area)?

Are set-up instructions changed when required? ❐ *yes* ❐ *no*

If yes, complete the following:

Who (job title) is responsible for approving the change?

Where is the updated set-up stored?

Who (job title) is responsible for maintaining the set-up instructions?

In-process Inspection Records

Is there a form in which the inspector records the inspection results? ❒*yes* ❒ *no*

If yes, complete the following:

What is the name of the report (i.e., History Card, Receiving Report)?

What is recorded onto the report (quantity passed, dimensions)?

Who (job title) approves the report?

Records must identify the inspection authority for product release.

Where is the report stored?

How long is the report kept?

Is the inspection status indicated on the item? ❒*yes* ❒ *no*

If yes, what marks are applied (i.e., stamp, tag, label)?

List the defect prevention method (i.e., SPC, error-proofing, visual control) that is used on each production line?

Line	Method

Final Testing

Written Procedures

Is final test performed in accordance with a quality plan? ❒*yes* ❒ *no*

You must document final inspection and test procedures.

Do you have final test procedures? ❒*yes* ❒ *no*

If yes, complete the following:

Do the procedures include a checklist of test items? ❒*yes* ❒ *no*

Who is responsible for writing the procedures?

Are the procedures under document control? ❒*yes* ❒ *no*

Final Test and/or Final Calibration

Who (job title) is responsible for holding parts until the required final test is complete?

Who (job title) is responsible for final test?

What is the scope of the inspection (i.e., 100%, sample)?

Is the unit calibrated at final test? ❒*yes* ❒ *no*

If yes, complete the following:

Who (job title) performs the calibration?

Are computer printouts of test results stored with the records? ❒*yes* ❒ *no*

Test Results

How is the test result status identified on the unit (i.e., traveler, stamp, label, computer printout)?

Who (job title/department, i.e., Quality, Supervisor) approves the final test?

Is a form completed indicating the final test results? ❐ *yes* ❐ *no*

If yes, complete the following:

What is the name of the record?

Who (job title) is responsible for completing the record?

Records must identify the inspection authority for product release.

Where is the record stored?

How long is it kept?

Defective Units

Where are defective units held?

Are defective units recalibrated and retested? ❒yes ❒ no

Is a report completed? ❒yes ❒ no

If yes, complete the following:

Where is the report stored?

How long is the report kept?

Who (job title/department, i.e., Quality, Engineering) dispositions the defective unit?

Final Inspection

Written Procedures

Do you have final inspection procedures? ❒yes ❒ no

You must have final inspection procedures to verify in-process inspections.

Do the procedures include verification of in-process inspections and tests? ❒yes ❒ no

If yes, complete the following:

Who (job title) is responsible for writing the procedures?

Are the procedures under document control?

Do you have a checklist of inspection items? ❒yes ❒ no

What is included in the checklist?

verify inclusion of packing list items	❐ *yes*	❐ *no*
verify cosmetics	❐ *yes*	❐ *no*
verify workmanship	❐ *yes*	❐ *no*
verify labeling on box	❐ *yes*	❐ *no*
verify labeling on unit	❐ *yes*	❐ *no*
other		

Final Inspection

You must hold product until final inspections and tests are completed and authorized.

Who (job title) is responsible for holding parts until the required inspection is complete?

Who (job title) is responsible for the inspection?

Is the unit inspected against assembly drawings for visual, functional, and dimensional inspection? ❐ *yes* ❐ *no*

What is the scope of the inspection (i.e., 100%, sample)?

Final Inspection Results

How is the inspection status identified on the unit (i.e., traveler, stamp, label, computer printout)?

You must maintain records of inspections and tests.

Who (job title/department, i.e., Quality, Supervisor) approves the final inspection?

Is a form completed indicating the final inspection results? ❒*yes* ❒ *no*

If yes, complete the following:

What is the title of the form?

Who (job title) is responsible for completing the form?

Where is the form stored?

How long is it kept?

Layout Inspection and Functional Testing

Do your customers require layout inspection and functional verification? ❒yes ❒ no

You must perform layout inspection according to customer requirements.

If yes, is there a procedure(s) for the inspection? ❒yes ❒ no

What is the title of the form used to record the results?

Where are the results maintained?

Are the results available for customer review? ❒yes ❒ no

Training

You must ensure that verification activities are performed by trained personnel.

Is training provided? ❐ *yes* ❐ *no*

If yes, complete the following:

Who provides the training (i.e., Quality, certified trainers)?

Are training records maintained? ❐ *yes* ❐ *no*

If yes, complete the following:

Where are the records stored?

Who (job title) is responsible for maintaining the records?

How long are the training records kept?

Written Procedures and Related Records

List records or forms, with their corresponding part numbers, used in this procedure:

List related work instructions and procedures, with their corresponding part numbers, that employees use as instructions for activities described above:

INSPECTION AND TESTING PROCEDURE TEMPLATE

1. PURPOSE

- *To establish close control on the quality level of all procured parts, materials, and services.*
- *To inspect, test, and identify product according to a documented plan.*
- *To identify nonconforming product.*
- *To ensure that inspection and tests have been performed to meet compliance to all contract specifications and drawing requirements.*
- To ensure that inspection and testing activities are prevention-based.
- To ensure that the incoming inspection system reviews subcontractor inspection data when determining the amount of incoming inspection to perform.
- To conduct layout inspection according to customer requirements.

2. SCOPE

This procedure applies to all parts, components, and materials received from suppliers or manufactured by the company and used in the manufacture, assembly, and shipment of products. This procedure covers all work or rework performed by an outside source.

3. RESPONSIBILITIES

<Job title> trains the operators to perform the tests.

<Job title> ensures that the quality levels required by contract specification are in place and that compliance to documentation is met.

<Job title> develops inspection and test plans to inspect parts at certain points in the operation.

<Job title> inspects work at the completion of an operation.

<Job title> conducts layout inspections.

4. PROCEDURE

4.1 Test Verification

See Inspection, Measuring, and Test Equipment, xxx.

<Job title> is responsible for maintaining the necessary gages, tools, meters, instruments, and equipment in the calibration system. Calibration records are stored in <location>.

4.2 Receiving - General

see Inspection and Test Status, xxx.

(Sample plans) are inspected.

<Company> requires the supplier by contract to supply inspection records according to <Company's> requirements.

<Job title> reviews the inspection data. <Job title> approves the inspection data. The status of the inspection data is marked using <markings>.

The reviewed inspection data are stored <location>.

Parts ordered from subcontractors certified for ship-to-stock do not require incoming inspection.

If applicable, the supplier identifies in advance nonconforming material and makes special arrangements.

4.3 Incoming Inspection

See Purchasing, xxx.

<Job title> performs incoming inspections. <Job title> is responsible for holding parts until the required inspection and tests are complete.

The inspector verifies the received product using <documentation> and follows <name of written procedure, work instruction, control plan> for the test and inspection.

The inspector records the inspection results onto <name of report> and indicates <items>. <Job title> approves the report. <Job title> stores the <name of report> in <location> for <length of time>.

<Job title> identifies the inspection status on the inspected item using <marking>.

4.4 Emergency Release

Product that is released for urgent production and that is not inspected is identified by <marking>. <Job title> records the non-inspected part onto <name of form> The <name of form> is stored in <location>.

4.5 Incoming Inspection - Defective and Discrepant Parts

See *Control of Nonconforming Product*, xxx.

Inspection status is indicated by <marking>. Defective or discrepant parts are held <location>. <Job title> dispositions the defective item and completes the <name of form> to initiate the disposition.

4.6 In-process Testing

See Statistical Techniques, xxx.

In-process tests are performed following <name of written procedures, work instruction, control plan>. *The procedures include a checklist of inspection items and are under document control. <Job title> is responsible for ensuring test procedures are current.*

<Sample plan for attributes and variable> is tested. The results are measured using Statistical Process Controls (SPC).

The following lists the in-process tests:

Test	Description	Test Record

<Job title> records test results into <database, record>. <Job title> indicates the test status on the unit using <marking>.

4.7 In-process Inspections

See Quality Records, xxx.

<Job title> is responsible for holding parts until the required inspection and tests are complete.

<Job title> inspects <sample, quantity> in-process parts for <list>.

<Job title> inspects that the set-up piece matches the specifications. <Job title> approves the set-up piece and signs the <name of form>.

If the set-up piece is rejected, the following occurs: <activity>.

When required, set-up instructions are modified. <Job title> is responsible for approving the change. The updated set-up is stored <location>. <Job title> is responsible for maintaining the set-up instructions.

The following lists the inspections performed:

Inspection	Description	Inspection Record

<Job title> inspects parts according to <name of written procedure>.

The inspector records the inspection results onto <name of report> and indicates <items>. <Job title> approves the report. <Job title> stores the <name of report> in <location> for <length of time>.

4.8 Final Testing

Final tests are performed according to <name of procedure> and in accordance with the quality plan. The procedures include a checklist with test items.

<Job title> is responsible for writing the test procedures. The test procedures are under Document Control.

<Job title> is responsible for holding parts until the required final test is complete.

<Job title> calibrates the unit at final test according to <sample plan>. <Job title> records calibration results onto <name of record>.

4.9 Final Test Results

<Job title> approves the final test results.

<Job title> stores final test records in <location> for <length of time>.

<Job title> applies <marking> to the unit indicating the final test result.

Defective units are held <location>. <Job title> is responsible for recalibrating and retesting the defective unit. <Job title> completes the <name of report>.

<Job title> stores the <name of report> in <location> for <length of time>.

4.10 Final Inspection

<Job title> is responsible for generating final inspection procedures that are under Document Control.

The final test procedure incorporates a checklist that includes the following:

- *<Item>*
- *<Item>*
- *<Item>*

<Job title> is responsible for holding parts until the required inspection is complete. <Job title> inspects <sample plan> of units against <documentation> for <parameters>.

<Job title> approves the final inspection.

The inspection status is identified on the unit using <marking>.

<Job title> completes the <name of form> indicating the final inspection status.

<Job title> stores the <name of form> in <location> for <length of time>.

4.11 Layout Inspection

<Job title> conducts layout inspection per customer-specific requirements <frequency>. Results of layout inspections are recorded on <record, log> and maintained by <job title>.

4.12 Training

See *Control of Nonconforming Product, xxx.*

Inspectors receive training to verify processes and products. Training records are stored <location> for <length of time>.

5. RELATED DOCUMENTS

<inspection procedures>
<inspection data sheet>
<reject/accept tag>
<customer-specific layout procedures>
<control plans>
<laboratory accreditation>
<sampling plans>

Document and Data Control
Inspection, Test, and Measuring Equipment
Inspection and Test Status
Control of Nonconforming Product
Corrective and Preventive Action
Quality Records
Statistical Technique
Purchasing

<list work instructions>

Notes:

Inspection, Measuring, and Test Equipment

4.11

What is the job title and name of the person responsible for this procedure?

ISO 9000 Standard:

4.11 Control of inspection, measuring, and test equipment

4.11.1 General

The supplier shall establish and maintain documented procedures to control, calibrate, and maintain inspection, measuring, and test equipment (including test software) used by the supplier to demonstrate the conformance of product to the specified requirements. Inspection, measuring, and test equipment shall be used in a manner which ensures that the measurement uncertainty is known and is consistent with the required measurement capability.

Where test software or comparative references such as test hardware are used as suitable forms of inspection, they shall be checked to prove that they are capable of verifying the acceptability of product, prior to release for use during production, installation, or servicing, and shall be rechecked at prescribed intervals. The supplier shall establish the extent and frequency of such checks and shall maintain records as evidence of control (see 4.16).

Where the availability of technical data pertaining to the measurement equipment is a specified requirement, such data shall be made available, when required by the customer or customer's representative, for verification that the measuring equipment is functionally adequate.

NOTE 17 For the purposes of this American National Standard, the term "measuring equipment" includes measurement devices.

4.11.2 Control procedure

The supplier shall:

a) determine the measurements to be made and the accuracy required, and select the appropriate inspection, measuring, and test equipment that is capable of the necessary accuracy and precision;
b) identify all inspection, measuring, and test equipment that can affect product quality, and calibrate and adjust them at prescribed intervals, or prior to use,

against certified equipment having a known valid relationship to internationally or nationally recognized standards. Where no such standards exist, the basis used for calibration shall be documented;

c) define the process employed for the calibration of inspection, measuring, and test equipment, including details of equipment type, unique identification, location, frequency of checks, check method, acceptance criteria, and the action to be taken when results are unsatisfactory;

d) identify inspection, measuring, and test equipment with a suitable indicator or approved identification record to show the calibration status;

e) maintain calibration records for inspection, measuring, and test equipment (see 4.16);

f) assess and document the validity of previous inspection and test results when inspection, measuring, and test equipment is found to be out of calibration;

g) ensure that the environmental conditions are suitable for the calibrations, inspections, measurements, and tests being carried out;

h) ensure that the handling, preservation, and storage of inspection, measuring, and test equipment is such that the accuracy and fitness for use are maintained;

i) safeguard inspection, measuring, and test facilities, including both test hardware and test software, from adjustments which would invalidate the calibration setting.

NOTE 18 The meteorological confirmation system for measuring equipment given in ISO 10012 may be used for guidance.

QS 9000 Interpretations and Supplemental Quality System Requirements

The QS 9000 supplements to ISO 9001, 4.11, "Control of Inspection, Measuring and Test Equipment," are:

- Inspection, Measuring, and Test Equipment Records
- Measurement Systems Analysis

These additions require the supplier to maintain specific records of calibration/verification activity and to use the Measurement Systems Analysis (MSA) Reference Manual for statistical studies of calibrated equipment.

Suggested Procedures

QS 9000:

- Measurement Systems Analysis

General

Outside Calibration Service

Is your measurement and test equipment calibrated by a commercial laboratory? ❐*yes* ❐ *no*

If yes, complete the following:

Do you require from the laboratory:

traceability to National Institute of Standard Technology	❐*yes*	❐ *no*
accuracy, stability, and range for calibrations performed	❐*yes*	❐ *no*
records of calibrations performed	❐*yes*	❐ *no*
written procedures for calibration of each piece of equipment calibrated	❐*yes*	❐ *no*
certified calibration report with each of piece of equipment calibrated	❐*yes*	❐ *no*
label on each piece of equipment	❐*yes*	❐ *no*
list reference or transfer standards used in calibration system	❐*yes*	❐ *no*
Who (job title) certifies the laboratory?		

In-house Calibration

Do you perform in-house calibrations? ❐*yes* ❐ *no*

If yes, complete the following:

Who (job title) performs the calibrations?

Who (job title) prepares the calibration procedures?

Calibration of equipment must follow documented procedures.

Are calibration procedures in the document control system? ❐*yes* ❐ *no*

If yes, list the titles and numbers of the calibration procedures.

Records

What is included in the calibration record?

item identification number	❐*yes*	❐ *no*
date of last calibration	❐*yes*	❐ *no*
date of next calibration	❐*yes*	❐ *no*
name of company performing the calibration	❐*yes*	❐ *no*
standards used to calibrate the equipment	❐*yes*	❐ *no*
certificate number traceable to National Bureau of Standards	❐*yes*	❐ *no*
environmental conditions under which the calibration was done	❐*yes*	❐ *no*
any adjustments required to establish calibration intervals	❐yes	❐ no
revisions following engineering changes	❐yes	❐ no
gage condition and actual readings as received for calibration/verification	❐yes	❐ no

Who (job title) stores the calibration record?

Where is the record stored?

How long are the records kept?

What records are stored (i.e., Calibration Record, Repair Record, Tooling History Card, Calibration Logbook)?

You must maintain records of customer notification regarding the shipment of suspect product.

Are customers notified if product shipped to them was tested or inspected with equipment that was not in calibration? ❒yes ❒ no

If yes, complete the following:

Who (job title) prepares the notification?

What is the name of the record?

How is the customer notified(phone call, overnight letter, e-mail)?

When (frequency) is the customer notified?

Standards

Do you perform statistical studies as outlined in the Measurement Systems Analysis (MSA) Reference Manual? ❒yes ❒ no

You must ensure that the measurement uncertainty is consistent with measurement capability.

If yes, complete the following:

What is the title of the procedure(s) for MSA guidance?

Are the studies performed for all measurement systems on approved control plans? ❒yes ❒ no

If a standard other than the MSA is used, do you obtain customer approval for the criteria? ❒yes ❒ no

Are measurement standards accurate relative to the intended use on the equipment? ❒*yes* ❒ *no*

If yes, what is the collective uncertainty of the standard (i.e., not to exceed 25% of the acceptable tolerance for each characteristic being calibrated)?

If not traceable to National Institute of Science and Technology standards or if no standard exists, is the calibration laboratory required to meet specially developed criteria? ❒*yes* ❒ *no*

You must calibrate equipment to a nationally recognized standard.

Do you have acceptance criteria which calibrated items must pass (i.e., four-to-one ratio)? ❒*yes* ❒ *no*

Identification

Master List

You must identify the inspection, measuring, and test equipment.

Do you maintain a list of all equipment in the calibration system? ❐*yes* ❐ *no*

If yes, complete the following:

What is the name of the list (i.e., Calibration Logbook, history files)?

What does the list include (i.e., identification number, serial number, location, date of last calibration, next calibration due date)?

Who (job title) is responsible for maintaining the master list?

Where is the master list stored?

Are employee-owned gages used in the facility? ❐yes ❐ no

If yes, how do you assure that those gages are included in the calibration system and on the master list?

Calibration Label

Equipment must show the calibration status.

Is a calibration label affixed to all equipment in the calibration system? ❐*yes* ❐ *no*

If yes, complete the following:

What does the label indicate (i.e. calibration date, the tester, the date of the next calibration, identification number)?

Are calibration labels nonremovable? ❐*yes* ❐ *no*

If it is impractical to apply the label to the item, where is the label attached (i.e., to the storage container, on the wall next to the item)?

For special applications, do labels note ranges that may be out-of-tolerance? ❐*yes* ❐ *no*

If yes, what is the name of the record indicating a special application?

Environmental and Handling Conditions

Environmental conditions must be suitable.

Is equipment calibrated in the same environment in which it is used? ❐*yes* ❐ *no*

If yes, complete the following:

For in-house calibrations, how do you monitor temperature and humidity?

When calibration results are obtained in an environment which departs from standard conditions, do you make compensating corrections? ❐*yes* ❐ *no*

Are temperature and humidity readings recorded? ❐*yes* ❐ *no*

If yes, complete the following:

Who (job title) records the readings?

Where are the readings stored?

Handling and storage conditions must ensure accuracy.

Is equipment calibrated by personnel trained to handle it to ensure its accuracy and fitness for use? ❐*yes* ❐ *no*

Schedule

Calibration Schedule

You must establish the frequency of inspections as evidence of control.

What is the basis for scheduling the calibration of items (i.e., historical records, frequency of use, manufacturer's specifications, prior calibrations, level of use of the test equipment, stability of the test equipment)?

What is the minimum interval between calibrations?

Is new equipment inspected and calibrated prior to use? ❒*yes* ❒ *no*

Adjustments to Schedule

Are intervals adjusted when necessary? ❒*yes* ❒ *no*

If yes, complete the following:

What is the basis for adjusting intervals (i.e., required accuracy, stability, purpose, and degree of usage, acceptable accuracy becomes out-of-specification)?

Do you extend the interval? ❒*yes* ❒ *no*

If yes, what is the measurement (i.e., four consecutive passes)?

Do you shorten the interval? ❒*yes* ❒ *no*

If yes, what is the measurement (i.e. two failures, the interval shortened by 50%, third failure-removed for repair)?

Adequacy

Review

Are the test equipment, measurement standards, and test software reviewed regularly to assess the calibration system? ❐*yes* ❐ *no*

You must ensure that test equipment is capable of the required measurement accuracy.

Do you make the technical data available to customers when requested? ❐*yes* ❐ *no*

If yes, complete the following:

What is the frequency of the review?

Who (job title) performs the review?

What is determined during the review?

calibration interval ❐*yes* ❐ *no*
test procedures adequate ❐*yes* ❐ *no*
other

Are calibration results compared to previous calibrations to evaluate the stability of the standard? ❐*yes* ❐ *no*

You must document the validity of previous inspections.

If yes, who (job title) compares the results?

Repair

For in-house calibrations, what is the procedure when an item is out-of-tolerance?

Is the item repaired? ❐*yes* ❐ *no*

If yes, who (job title) repairs the item?

Are prior inspection records reviewed to determine the validity of the previous inspection? ❐*yes* ❐ *no*

To whom (job title, department) are failures reported?

Is a decision made to reinspect all products calibrated with the faulty test equipment or measuring device? ❐*yes* ❐ *no*

Training

Management must ensure that verification activities are performed by trained personnel.

Is training provided? ❏ *yes* ❏ *no*

If yes, complete the following:

Are training records maintained? ❏ *yes* ❏ *no*

Where are the records stored?

Who (job title) is responsible for maintaining the records?

Written Procedures and Related Records

List records or forms, with their corresponding part numbers, used in this procedure:

List related work instructions and procedures, with their corresponding part numbers, that employees use as instructions for activities described above:

INSPECTION, MEASURING, AND TEST EQUIPMENT PROCEDURE TEMPLATE

1. PURPOSE

- *To establish procedures in the calibration and handling of measurement and test equipment used in to verify conformance of products to the established quality requirements and specifications.*
- *To establish a system by which gauges and measuring equipment are controlled, calibrated, and maintained.*
- *To establish an administrative system that assures conformance of all tooling to specified requirements.*
- To perform statistical studies of test and inspection equipment.
- To maintain records of calibration and verification activities to ensure the adequacy of the product.

2. SCOPE

This procedure governs all special tools, gauges, instruments, fixtures, and testing devices used for manufacture, assembly, inspection, and test, as well as tooling used as a measure of acceptance criteria, including company- and employee-owned equipment. This equipment is:

- *Mechanical tools and gauges used to measure physical dimensional parameters and includes, but is not limited to, jigs, fixtures, mechanical hand tools, and gauges.*
- *Electronic test equipment, which includes but is not limited to, instruments for measuring and recording ac or dc voltage, ac or dc current, resistance, inductance, capacity, and frequency.*
- *Physical tools and gauges used to check physical parameters, such as mass, temperature, pressure, humidity, and all special tools, gauges, instruments, fixtures, and testing devices used for manufacture, assembly, and inspection, and test.*
- *Optical test equipment used to measure optical parameters and includes, but is not limited to, jigs, fixtures, optical gauges, and optical test stations.*

3. RESPONSIBILITIES

All employees using measuring and test equipment are responsible for seeing that an item of equipment is not used when its calibration period has expired and to return such items for calibration.

The calibration laboratory notifies <Company> of any measuring test equipment found to be out-of-tolerance (as stated by the manufacturer) and provides a copy of the calibration data highlighting the out-of-tolerance measurements.

<Job title> maintains records for each item scheduled for calibration and repairs out-of-tolerance equipment.

<Job title> is responsible for the coordination of retrieving and assigning equipment being serviced by a contracted calibration lab.

<Job title> audits and reviews the calibration procedures being performed and keeps the calibration records on file.

<Job title> approves all repairs to equipment by contractors outside of <Company> prior to issuing any purchase order.

<Job title> investigates the high frequency of out-of-tolerance equipment.

<Job title> informs customer of suspect or nonconforming product that was shipped to the customer after being inspected/tested with equipment not in calibration.

<Job title> ensures that statistical studies and acceptance criteria for calibrated equipment is performed per the guidance of the Measurement Systems Analysis Reference Manual.

4. PROCEDURE

4.1 General

Measurement and test equipment are calibrated by a commercial laboratory. The laboratory must:

- *<Item>*
- *<Item>*
- *<Item>*

<Job title> certifies the laboratory.

<Job title> performs in-house calibrations. Personnel calibrating equipment are qualified. Records of qualifications are maintained in <location>.

<Job title> prepares the calibration procedures for in-house calibration. Calibration procedures are maintained by document control.

4.2 Records

See Corrective and Preventive Action, xxx.

The calibration record includes:

- *<Item>*
- *<Item>*
- *<Item>*

<Job title> stores the <name of records> in <location> for <length of time>.

<Job title> notifies the customer of product that was shipped after it was inspected with equipment not in calibration. <Notification> to the customer occurs <frequency> and the problem is resolved in accordance with the Corrective and Preventive Action Procedure, xxx.

4.3 Standards

Measurement standards are accurate relative to the intended use of the equipment. The collective uncertainty of the standard does not exceed <n%> of the acceptable tolerance for each characteristic being calibrated.

If not traceable to National Institute of Science and Technology standards, or if no standard exists, the calibration laboratory is required to develop criteria.

Calibrated items must pass an acceptance criteria ratio of <ratio>.

Statistical studies analyzing the variation of the equipment are performed by <job title>. The acceptance criteria and methods for analysis are performed according to the guidance of the Measurement Systems Analysis Reference Manual.

4.4 Identification

<Job title> maintains a master list of all equipment in the calibration system. The <name of master list> includes <items>.

<Job title> stores the <name of master list> in <location> for <length of time>.

A non-removable calibration label is affixed to all equipment in the calibration system. The label indicates <items>.

If it is impractical to apply the label to the item, it is attached <location>.

For special applications, labels note ranges that may be out-of-tolerance. <Name of record> indicates special applications.

4.5. Environmental

Equipment is calibrated in the same environment in which it is used. For in-house calibrations, temperature and humidity are monitored <activity>.

When calibration results are obtained in an environment that departs from standard conditions, compensating corrections are made.

<Job title> records temperature and humidity readings and stores the readings <location>.

4.6 Schedule

Items are scheduled for calibration based on <activity>.

The minimum interval between calibrations is <interval>. New equipment is inspected and calibrated prior to use.

Intervals are adjusted when necessary based on <activity>. The interval is extended when <criteria>. The interval is shortened when <criteria>.

4.7 Adequacy

<Job title> reviews test equipment and measurement standards <frequency> to assess the calibration system. The following is determined during the review:

- *<Item>*
- *<Item>*

<Job title> compares calibration results to previous calibrations to evaluate the stability of the standard.

<Job title> records gage conditions and actual readings of the equipment prior to calibration/verification.

For in-house calibrations, items found to be out-of-tolerance are <activity>. <Job title> repairs the item.

Prior inspection records are reviewed to determine the validity of the previous inspection.

Failures are reported to <job title> who determines whether to reinspect all product calibrated with the faulty test equipment or measuring device.

5. RELATED DOCUMENTS

<calibration logbook>
<calibration record>
<audit of external calibration facility>

<customer notification of suspect product form>

Document and Data Control
Process Control
Inspection and Testing
Inspection and Test Status
Quality Records
Corrective and Preventive Action
Measurement Systems Analysis Reference Manual

<list work instructions>

Inspection and Test Status

4.12

What is the job title and name of the person responsible for this procedure?

ISO 9000 Standard

4.12 Inspection and test status

The inspection and test status of product shall be identified by suitable means, which indicate the conformance or nonconformance of product with regard to inspection and tests performed. The identification of inspection and test status shall be maintained, as defined in the quality plan and/or documented procedures, throughout production, installation, and servicing of the product to ensure that only product that has passed the required inspections and tests [or released under an authorized concession (see 4.13.2)] is dispatched, used, or installed.

QS 9000 Interpretations and Supplemental Quality System Requirements

The QS 9000 supplements to ISO 9001, 4.12, "Inspection and Test Status," are:

- Product Location
- Supplemental Verification

Product Location gives guidance to the supplier on the ISO 9001 term, "suitable means." For the automotive application, "suitable means" must consider more than product location. The inspection and test status needs to be identified for all product unless the product is in a controlled, maintained environment, such as an automated paint line or welding process.

Secondly, the supplier must comply with additional customer-specific verification or identification requirements.

Suggested Procedures

QS 9000:

- Customer-Specific Verification of Product

General

What is the quantity of testing (i.e., 100%, sample)?

Are all parts tested identified with the test status?

How is the status identified (i.e., label, stamp, tags, barcode scanned)?

Is the marking of the test status included in the procedure, either in the work instructions or through training? ❐*yes* ❐ *no*

Receiving Status

What markings are used to indicate the receiving status of a part (i.e., failure tag)?

Who (job title) is responsible for applying the marking to received items?

You must identify the inspection status.

Did training for receiving inspection include marking the status of the part? ❐*yes* ❐ *no*

Do receiving inspection procedures include marking the status? ❐*yes* ❐ *no*

Who (job title) is responsible for ensuring that only parts that pass the receiving inspection are used?

You must record the authority for release of incoming product.

Are any reports completed indicating the inspection status? ❐*yes* ❐ *no*

If yes, complete the following:

What is the title of the report (i.e., Receiving Inspection Report)?

Who (job title) is responsible for completing the report?

Who (job title) is responsible for maintaining the report?

Where is the report stored?

How long is the report kept?

In-process Status

You must identify the inspection status.

What markings are used to indicate the status of an in-process inspection or test (i.e., failure tag, operator numbers, signed routings, barcode scanner)?

Who (job title) is responsible for applying the marking to the inspected item?

Did training for in-process inspection and test include marking the status of the part? ❒ *yes* ❒ *no*

Do work instructions include marking the status? ❒ *yes* ❒ *no*

You must record the authority for release of in-process product.

Who (job title) is responsible for ensuring that only parts that pass in-process inspections move to the next operation?

Are any reports completed indicating the inspection status ? ❒ *yes* ❒ *no*

If yes, complete the following:

What is the title of the report (i.e., history card, Inspection Report)?

Who (job title) is responsible for completing the report?

Who (job title) is responsible for maintaining the report?

Where is the report stored?

How long is the report kept?

Final Status

What markings are used to indicate the final status of a part (i.e., QA label, computer printout)?

You must identify the inspection status.

Who (job title) is responsible for applying the marking to final inspection and test items?

Did training for final test and inspection include marking the status of the part? ❐ *yes* ❐ *no*

Do final inspection and test procedures include marking the status? ❐ *yes* ❐ *no*

Who (job title) is responsible for ensuring that only parts that passed the final inspection and test are shipped?

You must identify the authority for release of product.

Are any reports completed indicating the final status? ❐ *yes* ❐ *no*

If yes, complete the following:

What is the title of the report (i.e., history card)?

Who (job title) is responsible for completing the report?

Who (job title) is responsible for maintaining the report?

Where is the report stored?

How long is the report kept?

Installation Status

You must identify the inspection status.

Do you install your product at the customer site? ❒*yes* ❒ *no*

If yes, complete the following:

What markings are used to indicate the status of an installation inspection (i.e., inspection sticker)?

Who (job title) is responsible for applying the marking to the inspected item?

Did training for installation include marking the status of parts? ❒*yes* ❒ *no*

Do written procedures or checklists include marking the status? ❒*yes* ❒ *no*

Are any reports completed indicating the installation status? ❒*yes* ❒ *no*

If yes, complete the following:

What is the title of the report (i.e., Service Report)?

Who (job title) is responsible for completing the report?

Who (job title) is responsible for maintaining the report?

Where is the report stored?

How long is the report kept?

Supplemental Verification

When required by the customer, do you comply with any additional verification requirements?
❒yes ❒ no

Written Procedures and Related Records

List records or forms, with their corresponding part numbers, used in this procedure:

List related work instructions and procedures, with their corresponding part numbers, that employees use as instructions for activities described above:

INSPECTION AND TEST STATUS PROCEDURE TEMPLATE

1. PURPOSE

- *To ensure the inspection and test status of all manufactured products indicated.*
- *To ensure only products that pass the required inspections and tests are used or dispatched.*
- *To ensure all components, subassemblies, and assemblies manufactured and received at <Company> are identified of current status.*

2. SCOPE

This procedure applies to all components, subassemblies, and assemblies manufactured and produced at <Company>.

3. RESPONSIBILITIES

<Job title> ensures that only units passed for dispatch leave the company.

<Job title> inspects all products according to checklists, test instructions, and customer-specific requirements.

<Job title> identifies all operations that require inspection and testing throughout the manufacturing process and develops test and inspection procedures at required operations.

<Job title> ensures that personnel who perform tests and inspections are trained according to documented procedures.

<Job title> maintains inspection and test records, as appropriate.

4. PROCEDURE

4.1 General

<Sample plan> of parts are tested. All parts tested are identified with the test status using <marking>. <Name of procedure> indicates the marking of test status.

4.2 Receiving Status

<Job title> applies <marking> to incoming items indicating the receiving status. Receiving procedures describe the marking applied to the received items.

<Job title> is responsible for ensuring that only parts that passed the receiving inspection are used. <Job title> completes the <name of report>.

<Job title> stores the <name of report> in <location> for <length of time>.

4.3 In-process Status

<Job title> applies <marking> to the part indicating the status of in-process tests. Written instructions describe marking the status. <Job title> is responsible for ensuring that only parts that passed move to the next operation.

<Job title> completes the <name of report>.

<Job title> stores the <name of report> in <location> for <length of time>.

4.4 Final Status

<Job title> applies <marking> to the part indicating the status of final tests. Written instructions describe marking the status. <Job title> is responsible for ensuring that only parts that passed are shipped.

<Job title> completes the <name of report>.

<Job title> stores the <name of report> in <location> for <length of time>.

4.5 Installation Status

For products installed at the customer's site, <job title> applies <marking> to the item indicating the status.

<Job title> completes the <name of report>.

<Job title> stores the <name of report> in <location> for <length of time>.

5. RELATED DOCUMENTS

<inspection data sheet>
<final system checklist>

Product Identification and Traceability
Process Control
Inspection and Testing
Control of Nonconforming Product
Inspection, Measuring, and Test Equipment
Quality Records
Servicing

<list work instructions>

Control of Nonconforming Product

4.13

What is the job title and name of the person responsible for this procedure?

ISO 9000 Standard:

4.13 Control of nonconforming product

4.13.1 General

The supplier shall establish and maintain documented procedures to ensure that product that does not conform to specified requirements is prevented from unintended use or installation. This control shall provide for identification, documentation, evaluation, segregation (when practical), disposition of nonconforming product, and for notification to the functions concerned.

4.13.2 Review and disposition of nonconforming product

The responsibility for review and authority for the disposition of nonconforming product shall be defined.

Nonconforming product shall be reviewed in accordance with documented procedures. It may be

a) reworked to meet the specified requirements,
b) accepted with or without repair by concession,
c) regraded for alternative applications, or
d) rejected or scrapped.

Where required by the contract, the proposed use or repair of product (see 4.13.2b) which does not conform to specified requirements shall be reported for concession to the customer or customer's representative. The description of the nonconformity that has been accepted, and of repairs, shall be recorded to denote the actual condition (see 4.16).

Repaired and/or reworked product shall be reinspected in accordance with the quality plan and/or documented procedures.

QS 9000 Interpretations and Supplemental Quality System Requirements

The QS 9000 supplements to ISO 9001, 4.13, "Control of Nonconforming Product," are:

- Suspect Product
- Control of Reworked Product
- Engineering Approved Product Authorization

These additions require the supplier to treat questionable product the same as if it were nonconforming. Rework must be performed by qualified personnel according to rework instructions. Nonconformance data must be analyzed, and efforts must be taken to reduce the nonconformances. The supplier must assure that product going to dealerships or service does not have any visible signs of rework. The supplier must review and consense on subcontractor requests to use material, product, or processes that differ from those that have successfully completed the Production Part Approval Process. The supplier must submit subcontractor requests, as well as requests for the supplier's product or process, to the customer for written authorization

Suggested Procedures:

- *Control of Nonconforming Product*
- *Rework*
- *Product Recall*
- *Return Authorization*

QS 9000:

- Subcontractor Request for Authorization
- Request for Customer Authorization
- Production Part Approval Process

Identify Discrepant Material

Are nonconforming/suspect materials removed from production lines? ❒*yes* ❒ *no*

If yes, complete the following:

Who (job title) can remove nonconforming/suspect material?

Is there an attempt to rework and reinspect the defective material? ❒*yes* ❒ *no*

If yes, complete the following:

Rework instructions must be accessible and used by qualified personnel.

Where (location) are rework instructions maintained?

Who (job title) reworks the material?

Who (job title) reinspects the material?

Are rework instructions controlled documents? ❒yes ❒ no

Is the rework recorded onto a form?

If yes, what is the title of the form?

Reworked Product for Service Applications

Is product provided for dealer maintenance or repair? ❒yes ❒ no

Service parts must not show any exterior signs of rework without customer approval.

If yes, complete the following:

Do rework instructions include criteria for exterior appearance? ❒yes ❒ no

Is service material inspected to assure there are no visible signs of rework? ❒yes ❒ no

Is prior customer approval obtained for items that show visible rework? ❒yes ❒ no

Identification

What identification is used to identify nonconforming/suspect material (i.e., sticker, tag)?

You must identify and document nonconform-ing material.

Who (inspector, any employee) marks defective items?

Is a report completed identifying the defective item? ❐ *yes* ❐ *no*

If yes, complete the following:

What is the title of the report (i.e., Material Review Report, Inspection Report, Discrepant Material Report)?

Who (job title) completes the report?

What is listed on the report (i.e., test results, disposition, s/n, defect code, quantity)?

Holding Area

Where are identified defective materials held?

You must segregate nonconform-ing material.

Work Area	Holding Area
receiving	
in-process	
final	
other	

Is the holding area

locked	❐ *yes*	❐ *no*
caged	❐ *yes*	❐ *no*
restricted access	❐ *yes*	❐ *no*
other		

Are items in the holding area logged? ❐ *yes* ❐ *no*

If yes, complete the following:

Who (job title) logs the items?

What is the title of the log?

Investigate

Classification

Who (job title) inspects holding areas?

Are items in holding areas classified? ❐ *yes* ❐ *no*

You must evaluate non-conforming material.

If yes, complete the following:

Who (job title) classifies the items?

Define the classifications you use:

Classification	Description
	(i.e., affects safety or performance, MRB)
	(i.e., unfit for sale, does not pass inspection)
	(i.e., can be reworked)

Changes to Products/Processes

You must obtain customer approval prior to changing your product or process.

If a product has completed production part approval, is customer approval obtained prior to changing a product or process? ❒yes ❒ no

Is the approval written? ❒yes ❒ no

What is the name of the form used to record the customer approval?

Who (job title) obtains customer approval?

Disposition Authority

You must authorize personnel to disposition nonconform-ing material.

Are defective materials dispositioned? ❒*yes* ❒ *no*

If yes, complete the following:

Who (job title, team, i.e., MRB) dispositions the material?

Who (job titles) comprise the team?

Does the team do the following?

prevent discrepant material from being used in product ❒*yes* ❒ *no*
coordinate corrective action teams ❒*yes* ❒ *no*
disposition discrepant material ❒*yes* ❒ *no*
other

How often does the team meet?

How are emergencies handled (i.e., meetings convene immediately, authorized member dispositions item)?

Disposition Types

Do you have procedures to disposition material? ❒ *yes* ❒ *no*

You must review nonconforming material according to a documented procedure.

If yes, complete the following:

What is the title of the procedure?

Is the procedure in your document control system?

Who (job title) maintains the procedure?

Complete the table with the possible dispositions:

Disposition	Description
(i.e., Use As Is)	*Parts are used exactly as they are with no anticipated effect on product quality.*
(i.e., Not Usable)	*Parts cannot be used in their present state. Scrap.*
(i.e., Rework)	*Parts are reworked and reinspected.*
(i.e., Hold for Evaluation)	Hold parts pending customer authorization if product/process is different than currently approved.
Other	

Are any forms completed indicating the disposition? ❒ *yes* ❒ *no*

If yes, complete the following:

Who (job title) completes the form?

What is the title of the form?

Who (job title) approves the rework, if required?

What happens to items that cannot be reworked?

Determine Cause of Defective Incoming Parts

For nonconforming/suspect incoming materials, who (job title) investigates the vendor?

Is the item returned to the vendor? ❐ *yes* ❐ *no*

If yes, complete the following:

Who (job title) contacts the vendor?

What form is completed indicating the discrepancy to the vendor?

Who (job title) reviews the vendor corrective action?

Determine Cause of Defective In-process Parts

For nonconforming/suspect in-process materials, who (job title) investigates the source?

What does the investigation analyze?

process	❐ *yes*	❐ *no*
operation	❐ *yes*	❐ *no*
quality records	❐ *yes*	❐ *no*
service records	❐ *yes*	❐ *no*
customer complaints	❐ *yes*	❐ *no*
inspection failure rates	❐ *yes*	❐ *no*
subcontractor changes	❐ *yes*	❐ *no*
other		

Nonconforming Data and Analysis

You must analyze and reduce nonconformities.

Is the amount of nonconforming product quantified? ❐ yes ❐ no

Is the cause(s) of the nonconformance(s) determined? ❐ yes ❐ no

Is a plan developed to reduce the number of nonconformities? ❐ yes ❐ no

If yes, does the plan prioritize efforts to reduce the nonconformities? ❐ yes ❐ no

Who (job title) prepares the plan?

Who (job title, team name) established the goals for reduction?

Who (job title) reviews the status of the plan?

Is the status of nonconformance reduction efforts tracked? ❐yes ❐ no

Who (job title, team name) tracks the status of these efforts?

Waivers

Do you use a method of waivers to release nonconforming/suspect product?

❐ *yes* ❐ *no*

If yes, complete the following:

Who (job title) authorizes the waiver?

Is the customer notified of the waiver? ❐ *yes* ❐ *no*

What records do you keep indicating that a product has been waived for release?

Where is the record stored?

How long is the record kept?

Who (job title) is responsible for maintaining the record?

Rework

Initiate Rework

If so dispositioned for rework, who (job title) defines the rework and retesting requirements?

You must re-inspect repaired or reworked product.

Is a rework form completed? ❐*yes* ❐ *no*

If yes, complete the following:

What is the title of the form?

Who (job title) approves the rework for internal rework?

Who (job title) approves the rework for external rework?

Internal Rework

Who (job title) approves the rework form?

Where is the reworked material moved to?

Who (job title) inspects reworked material?

Are records of the rework maintained? ❐ *yes* ❐ *no*

If yes, complete the following:

What is the title of the rework record?

Who (job title) is responsible for completing the record?

Where is the record stored?

How long is the record kept?

External Rework

What information is sent to the vendor for the rework (i.e., P.O., shipping memo, rework procedure, retest requirements)?

Who (job title) is responsible for generating the above information?

Customer Notification

If reworked product does not conform to requirements, is the customer notified? ❒yes ❒ no

If yes, complete the following:

Who (job title) is responsible for reporting the reworked product to the customer?

What is the name of the form used to document the actual condition of the reworked product?

Where is this form stored?

Who (job title) approves this form?

You must record the condition of an accepted non-conformity.

Engineering Approved Product Authorization

Once a product has completed the Production Part Approval Process, are authorizations for changes to product and process obtained from the customer according to the customer specific change approval process? ❒yes ❒ no

If yes, complete the following:

Are subcontractor changes approved by the supplier prior to submitting the change for customer approval? ❒yes ❒ no

Are written records maintained of the customer approval? ❒yes ❒ no

Do the records track subcontractor changes? ❒yes ❒ no

Do the records indicate the expiration date of the authorization? ❒yes ❒ no

Do the records indicate the quantity of parts authorized under the change? ❒yes ❒ no

You must obtain customer approval if a process or product is different than a currently approved product or process.

You must assure that the product and process comply with specifications when the change authorization expires.

Is there a procedure to assure that the product and process comply with the original and superseding specifications when the authorization expires? ❒yes ❒ no

Is each shipping container marked with the authorization? ❒yes ❒ no

Is each shipping container marked per the customer's unique requirements? ❒yes ❒ no

Product Recall

Recalled Units

If recommended, who (job title) locates recalled units?

How are units located (i.e., database)?

How are customers and warehouse personnel notified of a recall?

What records are kept to indicate the recall of finished parts?

Who (job title) is responsible for maintaining the records?

Compliance Requirements

Must product pass regulatory and safety requirements? ❒*yes* ❒ *no*

If yes, which agencies stipulate these requirements?

Is finished product reviewed in the field to determine whether to recall the product based on safety and customer satisfaction? ❒*yes* ❒ *no*

If yes, complete the following:

Who (job title) performs the review?

What records are maintained?

Where are the records stored?

Return Authorization

Can a customer return product for evaluation and repair if necessary? ❒*yes* ❒ *no*

Is an authorization number issued? ❒*yes* ❒ *no*

Who (job title) issues the number?

Receipt of Returned Item

Who (job title) receives the returned item?

What records are kept to receive the item into your facility?

Evaluation

Is the item inspected to determine the problem, or verify the customer's complaint, or determine if the item is acceptable for inventory? ❑ *yes* ❑ *no*

If yes, complete the following:

Who (job title) inspects the returned item?

What is the title of the form which is completed?

Where is the completed form stored?

Is material dispositioned? ❑ *yes* ❑ *no*

If yes, complete the following:

Does the disposition follow a nonconforming/suspect product procedure? ❑ *yes* ❑ *no*

If not, describe the disposition used for returned items.

Customer Repair Authorization

Who (job title) notifies the customer of labor and material costs?

If the product is not under warranty, how is customer payment authorized (i.e., P.O. number, sales order)?

Repair of Returned Item

Who (job title) is responsible for the repair?

Who (job title) creates the work orders for the repair?

Where is the repair performed (assembly line, repair station)?

Is the item inspected prior to shipment? ❐ *yes* ❐ *no*

If yes, complete the following:

Who (job title) inspects the item?

What inspection records are kept?

Written Procedures and Related Records

List records or forms, with their corresponding part numbers, used in this procedure:

List related work instructions and procedures, with their corresponding part numbers, that employees use as instructions for activities described above:

CONTROL OF NONCONFORMING PRODUCT PROCEDURE TEMPLATE

1. PURPOSE

- *To evaluate defective products and products of questionable integrity.*
- *To evaluate the cause of defects and rework the part when possible.*
- *To create a permanent solution that prevents recurrence of problems.*

2. SCOPE

This procedure applies both to the review and subsequent disposition of suspect/defective material purchased for or assembled by <Company> and to suspect/defective products returned for repair or evaluation.

3. RESPONSIBILITIES

It is every employee's responsibility to identify nonconforming/suspect material as discrepant when it is noticed during normal work assignments.

<Job title> analyzes product deficiencies to determine their cause and establishes and tracks a prioritized reduction plan for nonconforming/suspect material and product.

<Job title> oversees the <team> activities, and handles discrepant materials from incoming inspection and repair, and:

- *Makes periodic sweeps of the production areas to remove any accumulation of discrepant material to the <location>.*
- *When requested, responds to unique situations where unexpected nonconforming/suspect material threatens to disrupt production and assists in the segregation of this material.*
- *Determines if an audit of stock or staging areas is necessary to head off additional nonconforming/suspect material from reaching production areas. If audits are necessary, removes any additional nonconforming/suspect material to the <location>.*
- *Dispositions material as to type of discrepancy. Returns minor non-conformities to production for rework. Discrepant vendor material is segregated on pallets or suitable containers. Scrap material is segregated from discrepant vendor material.*

<Job title> inspects incoming product for conformance and completes the <name of Inspection Report>.

<Job title> evaluates customer returns and determines if the product is to be repaired.

<Job title> ensures that nonconforming/suspect material does not accumulate in production areas, conducts or delegates authority for surveillance and purges, and notifies <department> if unexpected occurrences of nonconforming/suspect material are noticed.

<Job title> obtains written customer authorization to use a product or process if it is different than a customer-approved product or process.

4. PROCEDURE

4.1 Identify Discrepant Material

<Job title> is responsible for removing nonconforming/suspect materials from production lines.

<Job title> attempts to rework and reinspect the defective material. <Job title> reinspects the material. The rework is recorded onto <name of form>.

<Job title> marks defective items with <marking>. <Job title> completes the <name of report> indicating <items>.

4.2 Holding Area

The following identifies where defective materials are held:

Work Area	Holding Area
receiving	
in-process	
final	
other	

The holding area is restricted and <describe characteristics>.

<Job title> logs items in the holding area into <name of log>.

4.3. Investigate

<Job title> inspects holding areas and classifies items in the holding area as follows:

Classification	Description
	(i.e., affects safety or performance, MRB)
	(i.e., unfit for sale, does not pass inspection)
	(i.e., can be reworked)

<Job title/Team> dispositions defective materials following <name of procedure>.

<Frequency>, the <team>, composed of <job titles>, performs the following activities:

- *<Activity>*
- *<Activity>*
- *<Activity>*

To handle emergencies, <describe what happens>.

The following describes possible dispositions:

Disposition	Description
(i.e., Use As Is)	*Parts are used exactly as they are with no anticipated effect on product quality.*
(i.e., Not Usable)	*Parts can not be used in their present state. Scrap.*
(i.e., Rework)	*Parts are reworked and reinspected.*
(i.e., Hold for Evaluation)	Parts are awaiting customer authorization.
other	

<Job title> completes the <name of form> indicating the disposition. <Job title> approves the rework.

Rework is performed by <job titles or description>. Rework is performed in the <location>.

If an item cannot be reworked, the following occurs: <activity>.

<Job title> investigates the vendor upon receipt of defective incoming parts.

<Job title> contacts the vendor and completes the <name of form>. <Job title> reviews the vendor's corrective action.

<Job title> investigates nonconforming/suspect in-process materials and analyses:

- *<Activity>*
- *<Activity>*
- *<Activity>*

If necessary, a method of waivers is used to release nonconforming/suspect product. <Job title> authorizes the waiver.

<Job title> maintains the <name of record>, indicating the product waived for release, quantity authorized, and authorization expiration date. The <name of record> is stored in <location> for <length of time>.

4.4 Rework

If so dispositioned for rework, <job title> defines the rework and retesting requirements and completes the <name of form>.

<Job title> approves internal rework. <Job title> approves external rework.

For items that are intended for use by dealerships or other service providers, visible signs of external rework are not allowed. <Job title> inspects the reworked product to this criteria.

For external rework, <job title> sends <documentation> to the vendor.

For internal rework, the reworked material is moved to <location>. <Job title> inspects the reworked material and completes the <name of form>. <Job title> stores the <name of form> in <location> for <length of time>.

If the reworked product does not conform to requirements, <job title> notifies the customer and completes the <name of form>. <Job title> approves the <name of form>. The completed <name of form> is filed as a quality record in <location>.

4.5 Recall

If recommended, <job title> locates recalled units by reviewing <name of records>. Customers and inventory are notified of a recall by <activity>. <Name of record> indicates the recall of finished parts. <Job title> is responsible for maintaining the records.

<Product> must pass regulatory and safety requirements stipulated by <agencies>.

Finished product is reviewed in the field to determine whether to recall the product based on safety and customer satisfaction. <Job title> performs the review and completes the <name of form>.

4.6 Return Authorization

A customer can return product for evaluation and repair, if necessary, by contacting <department>.

<Job title> issues an authorization number.

<Job title> receives the returned item and enters the information onto <name of record> to indicate the receipt of the product into the facility.

<Job title> inspects the item to determine the problem, to verify the customer's complaint, and determine if the goods are acceptable for inventory. <Job title> completes the <name of form>.

<Job title> dispositions the returned item.

<Job title> notifies the customer of labor and material costs. If the product is not under warranty, the customer authorizes payment by <activity>.

<Job title> is responsible for the repair.

<Job title>creates the work orders for the repair.

The repair is performed at <location> and inspected prior to shipment.

<Job title> inspects the item and completes the <name of record>. The <name of record> is stored <location> for <length of time>.

4.7 Nonconforming Data Analysis

<Job title> collects data on nonconforming and suspect material and product. The information is analyzed in <name of report>. <Job title, Team name> develops a plan to reduce the amount of nonconformances and tracks progress toward meeting the goal established by <job title, team name>.

4.8 Customer Authorization of Product or Process

When a product or process is different than what the customer approved, <job title> obtains written customer authorization prior to the product or process being used. <Job title> reviews subcontractor requests for authorization and submits these requests to the customer for approval.

The authorization from the customer is maintained in <name of record> by <job title > for <length of time>. <Job title> assures that product and process used after the authorization expires conforms to the current specifications.

Material shipped on an authorization is marked according to customer requirements by <job title, department>.

5. RELATED DOCUMENTS

<deviation/waiver notice>
<engineering change notice (ECN)>
<rework form>
<accept/reject tag>
<customer authorization form>
<nonconforming data analysis report>

Document and Data Control
Purchasing
Process Control
Inspection and Testing
Corrective and Preventive Action
Servicing
Production Part Approval Process

<list work instructions>

Corrective and Preventive Action

4.14

What is the job title and name of the person responsible for this procedure?

ISO 9000 Standard:

4.14 Corrective and preventive action

4.14.1 General

The supplier shall establish and maintain documented procedures for implementing corrective and preventive action.

Any corrective or preventive action taken to eliminate the causes of actual or potential nonconformities shall be to a degree appropriate to the magnitude of problems and commensurate with the risks encountered.

The supplier shall implement and record any changes to the documented procedures resulting from corrective and preventive action.

4.14.2 Corrective action

The procedures for corrective action shall include:

a) the effective handling of customer complaints and reports of product nonconformities;
b) investigation of the cause of nonconformities relating to product, process, and quality system, and recording the results of the investigation (see 4.16);
c) determination of the corrective action needed to eliminate the cause of nonconformities;
d) application of controls to ensure that corrective action is taken and that it is effective.

4.14.3 Preventive action

The procedures for preventive action shall include:

a) the use of appropriate sources of information such as processes and work operations which affect product quality, concessions, audit results, quality records, service reports, and customer complaints to detect, analyze, and eliminate potential causes of nonconformities;

b) determination of the steps needed to deal with any problems requiring preventive action;

c) initiation of preventive action and application of controls to ensure that it is effective;

d) confirmation that relevant information on actions taken is submitted for management review (see 4.1.3).

QS 9000 Interpretations and Supplemental Quality System Requirements

The QS 9000 supplements to ISO 9001, 4.14, "Corrective and Preventive Action," are:

- Problem Solving Methods
- Returned Product Test/Analysis

These additions require the supplier to use a structured approach to problem solving and to analyze product returned from the customer.

Suggested Procedures:

- *Internal Corrective Action*
- *Supplier Corrective Action*
- *Preventive Action*

QS 9000:

- Structured Problem Solving

General

You document procedures for corrective and preventive action.

Do you have written corrective action procedures? ❐*yes* ❐ *no*

Do you have written preventive action procedures? ❐*yes* ❐ *no*

If yes, complete the following:

Who (job title) is responsible for writing the procedures?

Are the procedures under document control? ❐*yes* ❐ *no*

Are the procedures revised when changes to processes occur due to corrective or preventive actions? ❐*yes* ❐ *no*

You must use structured problem solving methods.

Do you have a procedure for structured problem solving (i.e., Ford's Team Oriented Problem Solving, Kepner Tregoe, Chrysler 7D)? ❐yes ❐ no

If yes, complete the following?

When is the problem solving method used (i.e., when the cause of a problem is unknown or determine root cause of a nonconformance to specification, customer request)?

What is the time frame for responding to a customer request?

What are the steps of the structured problem solving method?

Is there a form to record the status and results of the investigation? ❐yes ❐ no

What is the name of the form?

Corrective and Preventive Action System

Initiate Corrective Action

What types of discrepancies require corrective action (subcontractor nonconformance, parts that cannot be reworked, result of customer complaint)?

You must investigate the cause of nonconforming product.

Are parts returned from the customer analyzed for the cause of the problem? ❐yes ❐ no

You must analyze product returned from the customer.

Are records kept of the analyses? ❐yes ❐ no

Are the records available to the customer when requested? ❐yes ❐ no

Is corrective action taken to eliminate the root cause of the problem? ❐yes ❐ no

Who (job title, i.e., department, any employee, operators, Customer Service, Purchasing) can initiate a corrective action?

Is there a form to complete to initiate a corrective action? ❐*yes* ❐ *no*

If yes, complete the following:

What is the title of the form for product corrective action?

What is the title of the form for subcontractor corrective action?

Who (job title) is responsible for reviewing the form?

Who (job title) is responsible for assigning an employee to resolve the nonconformance?

Is there a time frame given to resolve the nonconformance? ❐*yes* ❐ *no*

If yes, what is the time frame?

Is a log maintained for tracking corrective actions? ❐ *yes* ❐ *no*

If yes, complete the following:

What is the title of the log?

Who (job title) is responsible for maintaining the log?

Where is the log stored?

Initiate Preventive Action

What information is evaluated to assess potential causes of nonconformity that require preventive action (processes, work operations, concessions, audit findings, quality records, service reports, customer complaints)?

You must eliminate potential causes of non-conforming product.

Who (job title, i.e., department, any employee, operators, Customer Service, Purchasing) can initiate a preventive action?

Is there a form to complete to initiate a preventive action? ❐ *yes* ❐ *no*

If yes, complete the following:

What is the title of the form to initiate a preventive action?

Who (job title) is responsible for reviewing the form?

Who (job title) is responsible for assigning an employee to assess solutions for preventive action?

Is there a time frame given to analyze a preventive action? ❐ *yes* ❐ *no*

If yes, what is the time frame?

Is a log maintained for tracking preventive actions? ❐ *yes* ❐ *no*

If yes, complete the following:

What is the title of the log?

Who (job title) is responsible for maintaining the log?

Where is the log stored?

Execute System

Who (job title, any employee) can offer solutions to a corrective or preventive action?

What is analyzed to eliminate potential causes?

You must analyze the cause or potential cause of non-conforming product.

processes	❐ *yes*	❐ *no*
work operations	❐ *yes*	❐ *no*
quality records	❐ *yes*	❐ *no*
service reports	❐ *yes*	❐ *no*
customer complaints	❐ *yes*	❐ *no*
other		

Is there a form for entering solutions? ❐ *yes* ❐ *no*

If yes, complete the following:

Is this the same form used for initiating the corrective or preventive action? ❐ *yes* ❐ *no*

If no, complete the following:

What is the title of the form?

Who (job title) is responsible for maintaining the form?

Where is the form stored?

Investigate Root Cause for Corrective Action

What is involved in determining the root cause of the nonconformance (i.e., communicate with employees, review tests results, review subcontractors)?

Does the assignee complete a report of the investigation and recommend solutions to prevent recurrence? ❒*yes* ❒ *no*

If yes, complete the following:

What is the title of the report?

Who (job titles, i.e., individual department managers, corrective action team) are responsible for reviewing the recommendation?

Who (job title) is responsible for ensuring that the corrective action instructions are implemented?

Determine Preventive Action Steps

What is reviewed to eliminate potential nonconformity (i.e., communicate with employees, review tests results, review subcontractors, review customer complaints, analyze service reports)?

Who (job title, department) analyzes information to assess the need for preventive action?

Does the evaluator complete a report of the analysis and recommend preventive action? ❒*yes* ❒ *no*

If yes, complete the following:

What is the title of the report?

Who (job titles, i.e., individual department managers, Preventive Action Team) are responsible for reviewing the recommendation?

Who (job title) is responsible for ensuring that the preventive action instructions are implemented?

Corrective Action Meetings

Are corrective actions reviewed at meetings? ❐ *yes* ❐ *no*

If yes, complete the following:

At what type of meeting is corrective action reviewed?

What is the frequency for corrective action meetings?

Who (job titles) attend corrective action meetings?

What is reviewed at the meetings (i.e., top five problems derived from statistical analysis, closed corrective actions, unresolved corrective actions, results of quality audits)?

Records

Are records of the meeting maintained?

If yes, complete the following:

What is the title of the record (i.e., meeting minutes)?

Where is the record stored?

Who (job title) is responsible for maintaining the record?

Preventive Action Meetings

Are preventive actions reviewed at meetings? ❐ *yes* ❐ *no*

If yes, complete the following:

At what type of meeting is preventive action reviewed?

What is the frequency for preventive action meetings?

Who (job titles) attend preventive action meetings?

What is reviewed at the meetings (i.e., top five problems derived from statistical analysis, closed corrective actions, unresolved corrective actions, results of quality audits, service reports, customer complaints)?

Records

Are records of the meeting maintained?

If yes, complete the following:

What is the title of the record (i.e., meeting minutes)?

Where is the record stored?

Who (job title) is responsible for maintaining the record?

Resolve Problem

How are corrective action solutions maintained (i.e., database, log)?

How are preventive action recommendations maintained (i.e., database, log)?

If a change to a process or document is required, what form is used to initiate the change?

What inspections or tests are used to assess the effectiveness of the preventive action?

Who (job title) handles disputes between the company and subcontractor?

Are problematic subcontractors removed from the approved subcontractor list? ❐yes ❐ no

Verification of Effectiveness

You must ensure that corrective actions and preventive actions are effective.

Are test data used to verify the effectiveness of the action? ❐*yes* ❐ *no*

If yes, complete the following:

Which test data are analyzed (inspection records for receiving, in-process, final)? Specify the names of the records.

Are statistical analyses performed to determine the effectiveness of the corrective and preventive actions? ❐*yes* ❐ *no*

If yes, what measurements are used (i.e., run-time charts, histograms, Pareto)?

Is the corrective action log and preventive action log examined by Management to determine any trends in defects? ❐*yes* ❐ *no*

If yes, complete the following:

What is the frequency of review?

Permanent Changes

Who (job title, any employee) can request a change to a design or process?

How are major changes to the design or process implemented (i.e., ECO)?

You must implement changes to procedures.

What does a permanent change to a design or process affect (work instructions, manufacturing processes, subcontractors)?

Written Procedures and Related Records

List records or forms, with their corresponding part numbers, used in this procedure:

List related work instructions and procedures, with their corresponding part numbers, that employees use as instructions for activities described above:

CORRECTIVE AND PREVENTIVE ACTION PROCEDURE TEMPLATE

1. PURPOSE

- *To establish and specify systematic steps for corrective action in the resolution of quality-related problems.*
- *To analyze and resolve quality problems.*
- *To create a permanent solution that prevents recurrence of nonconformance or potential nonconformance.*

2. SCOPE

This procedure applies to internal or external customers of any product offered by <Company> that is evaluated for quality, reliability, safety, or performance.

Any written or oral expression of dissatisfaction by either internal or external customers related to the identity, quality, reliability, safety, or performance of any product or service offered by <Company> is subject to investigation for root cause and irreversible corrective action.

3. RESPONSIBILITIES

<Job titles> review problems for resolution.

<Job titles> implement permanent changes through the <name of group or team>.

<Job title> maintains the log of corrective actions and solutions.

4. PROCEDURE

4.1 Written Procedures

<Job title> prepares the written procedure for corrective action activities. These procedures are maintained in the Document Control system and revised when changes to processes occur.

<Job title> prepares the written procedure for preventive action activities. These procedures are maintained in the Document Control system.

4.2 Request for Corrective Action

Corrective actions are required for supplier nonconformance, parts that cannot be reworked, parts returned from the customer, customer requests for corrective action, and quality audit findings.

For product corrective actions, <job title> initiates a corrective action request by completing the <name of form>. <Job title> reviews the <name

of form>. <Job title> assigns an employee to resolve the nonconformance within <time frame>.

For supplier corrective actions, <job title> initiates a corrective action request by completing the <name of form>. <Job title> reviews the <name of form>. <Job title> assigns an employee to resolve the nonconformance within <time frame>.

For corrective actions resulting from quality audit findings, <job title> initiates a corrective action request by completing the <name of form>. <Job title> reviews the <name of form>. <Job title> assigns an employee to resolve the nonconformance within <time frame>.

For corrective action resulting from customer requests or parts returned from the customer, <job title> initiates a corrective action request by completing the <name of form>. <Job title> reviews the <name of form>. <Job title> assigns an employee to resolve the nonconformance within <time frame>.

<Job title> maintains the <corrective action log>. The <corrective action log> is stored <location> for <length of time>.

4.3 Execute Corrective Action System

<Job title> analyzes <items> to eliminate potential causes of nonconformance.

<Job title> can offer solutions to a corrective action by completing the <name of form>.

If an extension is needed, a <name of form> is submitted to the originator of the corrective action stating the amount of additional time needed and reason for the extension. The request is filed with the <job title>.

The <job title> reviews the resolution and indicates acceptance on the <name of form>. If not acceptable, the <job title> may review and reassign the investigation. If the problem is not resolved it is brought to <team>. Upon approval, the <job title> ensures that the corrective action instructions are passed to appropriate personnel for implementation.

The <job title> indicates the corrective action closure on the <name of corrective action log>.

4.4 Investigate Root Cause

To determine the root cause, <job title> does the following:

- *<Activity>*
- *<Activity>*

- *<Activity>*

<Job title> completes a report of the investigation and recommends solutions to prevent recurrence on the <name of report>. <Job title> reviews the <name of report>. <Job title> is responsible for ensuring that the corrective action instructions are implemented.

4.5 Corrective Action Meetings

At <frequency> <name of meeting>, corrective actions are reviewed. <Job titles> attend the <name of meeting>. At the <name of meeting>, the following are reviewed:

- *<Item>*
- *<Item>*
- *<Item>*

<Job title> maintains records of the <name of meeting>. The <name of record> is stored in <location> for <length of time>.

4.6 Resolve Corrective Actions

Solutions to corrective actions are maintained in <name of log>.

If a change to a process or document is required, <name of form> is used to initiate the change.

<Job title> handles disputes between the company and vendor. Problematic vendors are removed from the approved vendor list.

4.7 Verification of Effectiveness

Test data from <names of tests> are used to verify the effectiveness of the corrective action.

Statistical analysis is performed to determine the effectiveness of the corrective action using <measurements>.

<Frequency>, Management examines the corrective action log to determine any trends in defects. If any trends exist, appropriate corrective action is taken.

4.8 Permanent Changes

<Job title> can request a change to a design or process. Major changes to the design or process are implemented through <activity>. A permanent change to a design or process affects <documentation>.

4.9 Preventive Action System

<Job title, Department> evaluates <activities> to assess potential causes of nonconformity.

If a potential nonconformity is determined, <job title> completes the <name of form>. <Job title> reviews the <name of form> and assigns a responsible person to assess a solution for preventive action.

<Time frame> is given for the resolution of a preventive action.

Preventive actions are logged in the <name of log>. The <name of log> is stored <location>.

5. RELATED DOCUMENTS

<corrective action request (CAR) form>
<corrective action data log>
<status report>
<meeting minutes>
<preventive action request (PAR) form>
<preventive action log>

<Structured Problem Solving Procedure>
Management Review
Document and Data Control
Purchasing
Control of Nonconforming Product
Internal Quality Audits

<list work instructions>

Notes:

Handling, Storage, Packaging, Preservation, and Delivery

4.15

What is the job title and name of the person responsible for this procedure?

ISO 9000 Standard:

4.15 Handling, storage, packaging, preservation, and delivery

4.15.1 General

The supplier shall establish and maintain documented procedures for handling, storage, packaging, preservation, and delivery of product.

4.15.2 Handling

The supplier shall provide methods of handling product that prevent damage or deterioration.

4.15.3 Storage

The supplier shall use designated storage areas or stock rooms to prevent damage or deterioration of product, pending use or delivery. Appropriate methods for authorizing receipt to and dispatch from such areas shall be stipulated.

In order to detect deterioration, the condition of product in stock shall be assessed at appropriate intervals.

4.15.4 Packaging

The supplier shall control packing, packaging, and marking processes (including materials used) to the extent necessary to ensure conformance to specified requirements.

4.15.5 Preservation

The supplier shall apply appropriate methods for preservation and segregation of product when the product is under the supplier's control.

4.15.6 Delivery

The supplier shall arrange for the protection of the quality of product after final inspection and test. Where contractually specified, this protection shall be extended to include delivery to destination.

QS 9000 Interpretations and Supplemental Quality System Requirements

The QS 9000 supplements to ISO 9001, 4.15, "Handling, Storage, Packaging, Preservation, and Delivery," are:

- Inventory
- Customer Packaging Standards
- Labeling
- Supplier Delivery Performance Monitoring
- Production Scheduling
- Shipment Notification System

These additions require the supplier to have an inventory management system, comply with customer packaging and labeling requirements, establish a system that supports 100% on-time delivery, track delivery performance, notify the customer of shipments through on-line advance shipment notification, and to have a back-up system in case the on-line notification system fails. Production scheduling must be order-driven.

Suggested Procedures:

- *Handling: Training and Equipment*
- *Special Handling, ESD Control*
- *Storing Materials and Parts*
- *Stockroom Issue*
- *Packaging*
- *Preservation*
- *Generating Shipping Documents*
- *Delivery*

QS 9000:

- Inventory Control
- Packaging (Customer-Specific)
- Labeling of Customer Product
- Stock Rotation
- Advance Shipment Notification
- Delivery Performance

General

Do you have procedures for:

You must maintain procedures for handling, storage, packaging, preservation, and delivery.

handling material	❒*yes*	❒ *no*
storing parts	❒*yes*	❒ *no*
packaging units	❒*yes*	❒ *no*
preserving units		
delivering finished goods	❒*yes*	❒ *no*

Handling

Training

You must handle material in a manner that prevents damage and deterioration.

Is training provided for handling? ❒*yes* ❒ *no*

If yes, complete the following:

What type of training is provided (ESD control, forklift, safety)?

Are training records maintained? ❒*yes* ❒ *no*

If yes, complete the following:

Where are the records stored?

Who (job title) is responsible for maintaining the records?

Material Handling

How are materials handled to prevent damage (i.e., use of correct pallet jacks, containers, conveyors, forklifts, dollies, tote bins)?

Do you have a policy for handling and disposing of all environmentally-sensitive and hazardous items? ❒yes ❒ no

If yes, how is this policy conveyed?

ESD Control

Is ESD control practiced? ❒*yes* ❒ *no*

If yes, what is used to control electro-static discharge (i.e., ESD mats, wrist straps, ankle straps, training)?

Storage

Identification

What is used to identify the contents and physical characteristics of stored items (i.e., barcoding, labels)?

Are labels applied to boxes for identification? ❒*yes* ❒ *no*

If yes, complete the following:

Who (job title) applies the labels to the box (i.e., receiving, supplier)?

Temperature Control

Are heat-sensitive items under temperature control? ❐*yes* ❐ *no*

If yes, complete the following:

How is the room's temperature monitored?

Is refrigeration provided when necessary? ❐*yes* ❐ *no*

You must store material to prevent damage and deterioration.

Shelf-life Control

Do you practice First-In, First-Out (FIFO)? ❐yes ❐ no

Do you have any items whose shelf-life requires monitoring? ❐*yes* ❐ *no*

If yes, complete the following:

What items require shelf-life control (i.e., paint, batteries, chemicals, adhesives)?

How is the shelf-life date monitored (i.e., inspections)?

If monitored through inspections, is there an inspection report? ❐*yes* ❐ *no*

Are labels with the shelf-life date attached to the item?

Who (job title) ensures that the shelf-life date has not been exceeded?

You must inspect material at appropriate intervals.

What is the process for dispositioning items whose shelf-life has expired?

Hazardous Materials

Do you store hazardous materials? ❒*yes* ❒ *no*

If yes, complete the following:

Describe the storage area.

How is the hazardous storage area identified?

How is the area restricted (i.e., locked, caged)?

Storage Access

You must provide designated storage areas.

Is the stockroom:

caged	❒*yes*	❒ *no*
restricted	❒*yes*	❒ *no*
posted as a designated area	❒*yes*	❒ *no*

Storage Maintenance

Who is responsible for maintaining the cleanliness of the stockroom area?

If bins are used to hold work cell inventory, who (job title) is responsible for maintaining cleanliness and quantity?

Storage Inspection

If a problem is found during manufacturing, are parts removed from stock and put on hold? ❒yes ❒ no

If yes, complete the following:

Who (job title) is responsible for notifying the stockroom of the discrepant materials?

How are items quarantined?

Control of Supplies

Requisitions

Are materials issued from a stockroom? ❒yes ❒ no

Is there a requisition form used for issuing materials? ❒yes ❒ no

You must authorize receipt for dispatch.

Is yes, complete the following:

What is the name of the form?

Who (job title) can authorize the requisition?

What records are maintained to control the inventory?

Who (job title) delivers the materials to the work area?

Assembling a Stock Pick

Is the pick assembled from the Bill of Materials? ❒yes ❒ no

Who (job title) is responsible for assembling the pick?

Who (job title) is responsible for distributing the pick to the work areas?

When an assembly (subassembly) is complete, is it stored in the stockroom?

Do any records accompany the completed assembly?

If yes, complete the following:

What is the title of the record?

Who (job title) completes the record?

Is the inventory record updated? ❒*yes* ❒ *no*

If yes, what is the name of the inventory record?

Who (job title) is responsible for updating the inventory record?

Just-In-Time (JIT) Inventory Control/Inventory Management System

Who (job title) is responsible for ensuring that each work center has adequate stock for the day/shift?

Describe what happens if additional stock is required for the day/shift (i.e., requisition completed, changeover stock issued or returned to stockroom)?

You must have an inventory management system.

Do you have a documented inventory management system? ❒yes ❒ no

If yes, complete the following:

Is it an automated system? ❒yes ❒ no

What is the name of the system?

Who (job title) is responsible for ensuring that the system is fully operable?

Are materials inspected prior to being packed in the work cell? ❒*yes* ❒ *no*

Are minimum and maximum quantities of parts determined? ❒*yes* ❒ *no*

If yes, complete the following:

Who (job title) determines the quantities?

What information is used to make this assessment (i.e., time budgets, work orders)?

Who (job title) approves the quantity plan?

Who (job title) is responsible for monitoring the quantity of parts in the work center?

Who (job title) moves parts to work centers?

If budgeted quantities need modification, who (job title) can make this change?

Inventory Records

What records are maintained to control the inventory and optimize inventory turns over time?

How do you ensure proper stock rotation?

How does your inventory management system minimize inventory levels?

Are records maintained to show the number of parts consumed? ❒*yes* ❒ *no*

If yes, complete the following:

What is the title of the report?

Who (job title) is responsible for completing the report?

How frequently is the report prepared?

Who (job title) reviews the report?

Cycle Count

Are cycle counts performed? ❒*yes* ❒ *no*

If yes, complete the following:

What items are counted (high movers, costly items)?

What is the frequency of the cycle count?

Who (job title) counts the inventory?

Who (job title) is responsible for adjusting the inventory list to match the cycle count?

Do you perform a physical inventory? ❒ *yes* ❒ *no*

If yes, complete the following:

Who (job title) is responsible for administering the physical inventory process?

Who (job title) performs the actual inventory count?

What records are maintained for physical inventory?

Who (job title) maintains the records?

Packaging

Instructions

You must pack product to ensure conformance to specified requirements.

Are packaging instructions available? ❐ *yes* ❐ *no*

If yes, complete the following:

Do instructions explain the methods of cleaning and preserving (including moisture elimination), cushioning, blocking, and crating, if applicable? ❐ *yes* ❐ *no*

Who (job title) packages the completed unit (i.e., packaging operators, carpenters)?

Who (job title) is responsible for generating the packaging instructions?

Verification

Is inspection data verified before packaging? ❐ *yes* ❐ *no*

If yes, complete the following:

Who (job title) verifies the data?

How is the verification status indicated (i.e., sign and date form, QC stamp)?

Are contents verified before sealing the package? ❐ *yes* ❐ *no*

If yes, complete the following:

What documentation is used to verify contents (i.e., packing list, invoice)?

Who (job title) is responsible for verifying the contents?

What records are maintained to indicate that the contents were verified, if any?

Packaging Materials

Are protective materials used to package finished goods? ❑yes ❑ no

List some of the packing materials.

Does the customer specify packaging? ❑yes ❑ no

Do the Bill of Materials and prints call out materials for packaging (i.e., type of wood, shock mounts)? ❑yes ❑ no

Preservation

Preservation Through Packaging

Who (job title) designs your packaging to ensure proper preservation, if applicable?

You must use appropriate methods to preserve product.

Who (job title) certifies that your packaging preserves product, if applicable?

Who (job title) tests that packaging assures proper product preservation, if applicable?

If packaging fails, what is the corrective action?

Does your packaging meet National Safe Transit Association (NSTA) guidelines, if applicable? ❑yes ❑ no

Finished Product Segregation

You must use appropriate methods to segregate product.

Are markings and labels legible and durable and in the required language? ❒*yes* ❒ *no*

Are units identified on the package? ❒*yes* ❒ *no*

If yes, complete the following:

What are the identifying marks (i.e., serial number, P.O. number, part number)?

Are tamper-proof labels used? ❒*yes* ❒ *no*

Are quality inspection stamps or labels applied to the package? ❒*yes* ❒ *no*

Is documentation applied to the outside of the package? ❒*yes* ❒ *no*

If yes, complete the following:

What is the documentation affixed to the box (i.e., packing list, invoice, shipping papers)?

Are packages labeled with:

names and addresses ❒*yes* ❒ *no*
special handling instructions ❒*yes* ❒ *no*
shock watch indicators ❒*yes* ❒ *no*
tip and tell indicators ❒*yes* ❒ *no*
country-specific handling instructions ❒*yes* ❒ *no*
export documentation weights ❒*yes* ❒ *no*
customer-specific requirements ❒yes ❒ no
other

Delivery

Finished Goods Storage

You must protect product after final inspection and test.

Are packaged materials protected prior to delivery? ❐ *yes* ❐ *no*

If yes, is the area secured? ❐ *yes* ❐ *no*

Are deliveries F.O.B.? ❐ *yes* ❐ *no*

If yes, complete the following:

Does the customer specify the shipping requirements? ❐ *yes* ❐ *no*

Supplier Delivery Performance Monitoring

You must have a system that ensures 100% on-time delivery to the customer.

Does your delivery system meet 100% on-time customer delivery requirements? ❐ yes ❐ no

If yes, complete the following:

How do you monitor your delivery performance to assure 100% on-time delivery?

You must track your delivery performance.

Do you track the performance of the delivery system? ❐ yes ❐ no

Who (job title) tracks the performance of the delivery system?

Who (job title, team name) reviews the performance of the delivery system?

What action is taken if 100% on-time delivery is not met?

Do you inform the customer of the corrective action taken as a result of not meeting 100% on-time delivery? ❒yes ❒ no

If yes, how do you inform the customer of this action?

Do you adhere to customer established lead time requirements? ❒yes ❒ no

If yes, complete the following:

How do you evaluate and monitor your adherence to the lead time requirements (i.e., through tracking on-time delivery performance)

Who (job title) is responsible for reviewing customer delivery requirements?

How do you ensure that you have the most up-to date customer-specified transportation mode, routings, and containers?

Who (job title) is responsible for scheduling production?

For OEM or Service parts, is this scheduling order-driven? ❒yes ❒ no

Shipment Notification System

Do you have a computerized advance shipment notification for on-line transmittals?

❐yes ❐ no

If yes, complete the following:

Does this system transmit the required customer shipping information at the time of shipment? ❐yes ❐ no

Who (job title) is responsible for the system?

Does the system verify that all Advance Shipment Notifications (ASNs) match the shipping documents and labels? ❐yes ❐ no

Is there a back-up system in case the on-line system fails? ❐yes ❐ no

If yes, describe the system.

Does the back-up system verify that all ASNs match the shipping documents and labels?

❐yes ❐ no

Are records maintained to verify shipment according to customer requirements when the back-up system is used? ❐yes ❐ no

If yes, what is the name of the record?

Who (job title) maintains the record?

Records

Do you provide packing lists? ❑*yes* ❑ *no*

If yes, complete the following:

What is found on the packing list (i.e., customer information, part number, serial number, customer invoice number)?

Who (job title) is responsible for printing the packing list?

Is a Bill of Lading generated? ❑*yes* ❑ *no*

If yes, complete the following?

Who (job title) is responsible for printing the Bill of Lading?

How is the Bill of Lading generated?

Do you use a shipping log? ❑yes ❑ no

If yes, complete the following:

Who (job title) is responsible for maintaining the shipping log?

What information is found on the shipping log (i.e., delivery lot identification, shipping date, destination, inspector)?

How long is the shipping log maintained?

Where is the shipping log stored?

Vehicles

Does the customer specify the delivery vehicle? ❐ *yes* ❐ *no*

Does your company supply the delivery vehicle? ❐ *yes* ❐ *no*

If yes, complete the following:

Are delivery trucks safety regulated? ❐ *yes* ❐ *no*

If yes, complete the following:

What records are maintained to ensure safety regulations?

Where are the records stored?

Are delivery trucks inspected? ❐ *yes* ❐ *no*

If yes, complete the following:

Who (job title) is responsible for inspecting the vehicles?

What is the frequency of inspection?

Are records of inspection maintained? ❐ *yes* ❐ *no*

If yes, complete the following:

What is the title of the inspection record?

Who (job title) is responsible for the inspection record?

Where are inspection records stored?

Written Procedures and Related Records

List records or forms, with their corresponding part numbers, used in this procedure:

List related work instructions and procedures, with their corresponding part numbers, that employees use as instructions for activities described above:

HANDLING, STORAGE, PACKAGING, PRESERVATION, AND DELIVERY PROCEDURE TEMPLATE

1. PURPOSE

- To control the handling of products and parts from the point of receipt through storage, until they are sent to their final destination.
- *To ensure that material is moved or handled in a safe manner that prevents damage or deterioration.*
- *To provide secure storage areas to prevent damage or deterioration of product, pending use or delivery.*
- *To preserve product integrity and product identification through packaging materials.*
- *To ensure that the product is protected after final inspection and test, and where contractually applicable, to protect the product through delivery to the customer.*
- To establish a system for on-line notification of shipments to the customer
- To establish a system for monitoring delivery performance.
- To ensure that an inventory control system is established that optimizes inventory turns over time.

2. SCOPE

This procedure applies to all products and parts manufactured or sold by <Company>.

3. RESPONSIBILITIES

<Job title> packages product according to the work instructions.

<Job titles> handle goods through the manufacturing process and packages finished goods prior to shipment.

Delivery trucks and drivers are subcontracted through an approved vendor carrier.

The Environmental Health and Safety Manager sets policy for handling environmentally hazardous material.

<Job title> ensures that inventory is available at the work center.

<Job title> verifies deliveries from vendors and matches the purchase order and vendor packing list.

<Job title> picks the product from the warehouse and generates a Pick List.

<Job title> maintains the inventory and distributes stock to the work centers.

4. PROCEDURE

4.1 General

<Job title> generates and updates procedures for handling materials. These procedures are in the Document Control system.

<Job title> generates and updates procedures for storing parts and assemblies. These procedures are in the Document Control system.

<Job title> generates and updates procedures for packaging units. These procedures are in the Document Control system.

<Job title> generates and updates procedures for inventory management. These procedures are in the Document Control system.

<Job title> generates and updates procedures for Advance Shipment Notification (ASN). These procedures are in the Document Control system.

<Job title> generates and updates procedures for delivery of finished goods. These procedures are in the Document Control system.

4.2 Handling

Employees participate in training for handling, including:

- *<Course>*
- *<Course>*
- *<Course>*

<Job title> stores <training records> in <location> for <length of time>.

To prevent damage during material handling, the following devices may be used:

- *<Device>*
- *<Device>*
- *<Device>*

Disposal of environmentally sensitive material follows the <name of policy>. <Job title> conveys this policy through <activity>.

4.3 ESD Control

Control of Electro-Static Discharge (ESD) is ensured in the warehouse through the use of <ESD devices>. In the manufacturing area, ESD is controlled though the proper use of <ESD devices>.

<Job title> participates in <ESD training>. <Job title> maintains the training records.

4.4 Storage

<Markings> identify the contents and physical characteristics of stored items. <Job title> applies the <markings> according to work instructions.

4.5 Temperature Control

Heat-sensitive items are under temperature control. <Job title> monitors the room's temperature using <activity>. Refrigeration is provided when necessary.

4.6 Shelf Life Control

First-in, First-Out (FIFO) rotation is used. The following items require shelf life monitoring:

- *<Item>*
- *<Item>*
- *<Item>*

The items are dated with expiration date labels. <Job title> inspects the label <frequency>. <Job title> completes <name of inspection report>. Items with expired shelf-life are dispositioned by <job title>.

4.7 Hazardous Materials

Hazardous materials are stored in the <storage area>. <Identification> identifies the area as hazardous storage. The <storage area> is <characteristics>.

4.8 Storage Access

Access to the stockroom is restricted by:

- *<Attribute>*
- *<Attribute>*
- *<Attribute>*
- *<Attribute>*

4.9 Storage Maintenance

<Job title> maintains the cleanliness of the stockroom area. <Job title> maintains cleanliness and quantity of work cell bins.

4.10 Storage Inspection

If a problem is found during manufacturing, parts are removed from stock and put on hold. <Job title> notifies the stockroom of the discrepant materials. Items are quarantined in the following manner:

- *<Activity>*
- *<Activity>*

4.11 Control of Supplies

Upon receiving the Bill of Materials, <job title> assembles the pick from the stockroom. <Job title> distributes the pick to the work areas. Upon completion of an assembly or subassembly, the item and <documentation> is stored in the stockroom. <Job title> is responsible for maintaining the inventory record.

<Job title> is responsible for ensuring each work center has adequate stock for the day. If additional stock is required for the day, stock is issued. <Job title> completes the <name of form>.

<Job title> determines the minimum and maximum quantities stored in the work cell. <Items> are used to assess the quantity. <Job title> approves the quantity plan.

<Job title> is responsible for monitoring the quantity of parts in the work center. If budgeted quantities need modification, <job title> can make this change.

<Frequency>, <job title> completes the <name of inventory report> indicating the number of parts consumed. The report is reviewed by <job title>.

4.12 Just-In-Time (JIT) Inventory/Inventory Management System

A JIT material control system is used for work in process.

The materials are stored in <location>. Records on the <location> show the minimum and maximum levels. <Job title> is responsible for ensuring proper inventory levels.

<Job title> moves inventory to the required location.

4.13 Cycle Count

<Frequency>, <job title> performs a cycle count of <type of items>.

<Job title> is responsible for adjusting the inventory list to match the cycle count.

4.14 Packaging Instructions

Packaging instructions explain the methods of cleaning and preserving, (including moisture elimination), cushioning, blocking, and crating, if applicable. <Job title> packages the completed unit.

<Job title> generates the packaging instructions and assures that the product is packaged according to customer-specific requirements..

4.15 Packaging Verification

<Job title> verifies inspection data and label information, packages the unit, and applies <marking> indicating the verification status.

<Job title> verifies the contents before sealing the package against <documentation> and indicates the verification on the <name of record>.

4.16 Packaging Materials

Protective materials, such as <items>, are used to package finished goods.

Packaging is specified by the <customer, bill of materials>.

4.17 Packaging ID

Markings and labels are legible, durable, in the required language, and tamper-resistant.

Units are identified on the package using the following markings:

- *<Marking>*
- *<Marking>*
- *<Marking>*

<Job title> ensures that product is identified with the required customer-specific labeling.

Quality inspection stamps or labels are applied to the package. <Documentation> is applied to the outside of the package.

Packages are labeled with:

- *<Item>*

- *<Item>*
- *<Item>*
- *<Item>*

4.18 Preservation

<Job title> designs the packaging to ensure preservation.

<Job title> certifies the packaging.

<Job title> tests the packaging.

If packaging fails, corrective action is taken as follows: <activity>.

Packaging meets National Safe Transit Association (NSTA) guidelines.

4.19 Delivery

See Corrective and Preventive Action, xxx.

Finished goods awaiting delivery are stored <location>. This <location> is secured by means of <attribute>.

The customer specifies the shipping requirements and <job title> ensures that the delivery conforms to the customer's latest requirements, including mode of shipment, routings, and containers.

<Job title> issues the on-line ASN according to the customer's requirements.

<Job title> tracks delivery performance through:

- <Activity>
- <Activity>

When delivery performance is below 100% on-time to the customer, <job title> initiates corrective action according to the Corrective and Preventive Action Procedure.

<Job title> informs the customer by <activity> of late shipment information.

4.20 Records

<Job title> prints the packing list. The packing list includes <information>.

<Job title> generates the Bill of Lading.

<Job title> maintains the shipping log that includes <information> and the shipment notification system. The shipping log is stored <location> for <length of time>.

4.21 Vehicles

The customer specifies the delivery vehicle. Delivery trucks are safety regulated. <Job title> maintains records to ensure safety regulations are met in <location>.

<Frequency>, <job title> inspects the delivery vehicles and completes the <name of record>. The <name of record> is stored <location> for <length of time>.

5. RELATED DOCUMENTS

<inventory report>
<receiving report>
<issue transaction form>
<packing list>
<Bill of Lading>
<Advance Shipment Notification (ASN) database>

Process Control
Document and Data Control
Product Identification and Traceability
Inspection and Testing
Inspection and Test Status
Quality Records
Training
Corrective and Preventive Action

<list work instructions>

Control of Quality Records

4.16

What is the job title and name of the person responsible for this procedure?

ISO 9000 Standard:

4.16 Control of quality records

The supplier shall establish and maintain documented procedures for identification, collection, indexing, access, filing, storage, maintenance, and disposition of quality records.

Quality records shall be maintained to demonstrate conformance to specified requirements and the effective operation of the quality system. Pertinent quality records from the subcontractor shall be an element of these data.

All quality records shall be legible and shall be stored and retained in such a way that they are readily retrievable in facilities that provide a suitable environment to prevent damage or deterioration and to prevent loss. Retention times of quality records shall be established and recorded. Where agreed contractually, quality records shall be made available for evaluation by the customer or the customer's representative for an agreed period.

NOTE 19   Records may be in the form of any type of media, such as hard copy or electronic media.

QS 9000 Interpretations and Supplemental Quality System Requirements

The QS 9000 supplements to ISO 9001, 4.16, "Quality Records," are:

- Record Retention
- Superseded Parts

These additions require the supplier to maintain records according to the specific requirements in the QS 9000. The retention times in the QS 9000 are minimums and are not intended to replace government- or customer-specific requirements. Also, copies of documents from superseded parts that are needed for new part qualification must be retained in the new part file.

Storage

Where are Quality Records stored (i.e., in the individual departments responsible for the records, central location, database)?

You must have procedures to identify, collect, index, access, file, store, maintain, and dispose of Quality Records.

Is the length of time determined by individual departments? ❑*yes* ❑ *no*

How are Quality Records protected from damage?

How are Quality Records protected from loss?

How are Quality Records protected from deterioration?

Are Quality Records legible and readily retrievable? ❑*yes* ❑ *no*

Quality Records must be legible, identifiable to the product, readily retrievable, and prevented from loss or deterioration.

Do computerized records follow an established back-up procedure? ❑*yes* ❑ *no*

If yes, who (job title) is responsible for backing up the records?

Where (location) are documents from superseded parts retained ?

Automotive-specific Retention Requirements

You must retain certain records according to the QS 9000 requirements.

Are production part approval records maintained for a minimum of the length of time that the part is active for production and service plus one calendar year?

❑yes ❑ no

If no, what is the length of retention (i.e., x years due to government- or customer-specific requirement)?

Are tooling records maintained for a minimum of the length of time that the part is active for production and service plus one calendar year? ❒yes ❒ no

If no, what is the length of retention (i.e., x years due to government- or customer-specific requirement?

Are records of purchase orders and amendments maintained for a minimum of the length of time that the part is active for production and service plus one calendar year?
❒yes ❒ no

If no, what is the length of retention (i.e., x years due to government- or customer-specific requirement)?

Are quality performance records (i.e., control charts, inspection and test results) retained for a minimum of one calendar year after the year they were created?
❒yes ❒ no

If no, what is the length of retention (i.e., x years due to government- or customer-specific requirement)?

Are records of internal audits retained for a minimum of three years after audit completion?
❒yes ❒ no

If no, what is the length of retention (i.e., x years due to government- or customer-specific requirement)?

Are records of management reviews retained for a minimum of three years after the review?
❒yes ❒ no

If no, what is the length of retention (i.e., x years due to government- or customer-specific requirement)?

Records

Complete the following table for ***Inspection and Test Records:***

You must maintain records that demonstrate your Quality System. You must establish retention times.

Record Title	Description	Storage Area/Responsible Person (job title)	Length of Storage

Complete the following table for ***Traceability Data:***

Data Type/Title	Description	Storage Area/Responsible Person (job title)	Length of Storage

Complete the following table for ***Contract Review Reports:***

Report Title	Description	Storage Area/Responsible Person (job title)	Length of Storage

Complete the following table for ***Design Review Reports:***

Report Title	Description	Storage Area/Responsible Person (job title)	Length of Storage

Complete the following table for ***Training Records:***

Record Title	Description	Storage Area/Responsible Person (job title)	Length of Storage

Complete the following table for ***Audit Reports:***

Report Title	Description	Storage Area/Responsible Person (job title)	Length of Storage

Complete the following table for ***Nonconforming Product Reports:***

Report Title	Description	Storage Area/Responsible Person (job title)	Length of Storage

Complete the following table for ***Customer-Supplied Product Reports:***

Report Title	Description	Storage Area/Responsible Person (job title)	Length of Storage

Complete the following table for ***Calibration Reports:***

Report Title	Description	Storage Area/Responsible Person (job title)	Length of Storage

Complete the following table for ***Process Qualification Reports:***

Report Title	Description	Storage Area/Responsible Person (job title)	Length of Storage

Complete the following table for ***Equipment Verification Reports:***

Report Title	Description	Storage Area/Responsible Person (job title)	Length of Storage

Complete the following table for ***Management Review Reports:***

Report Title	Description	Storage Area/Responsible Person (job title)	Length of Storage

Complete the following table for ***Subcontractor Data:***

Data Type/Title	Description	Storage Area/Responsible Person (job title)	Length of Storage

Complete the following table for ***Corrective Action Reports:***

Report Title	Description	Storage Area/Responsible Person (job title)	Length of Storage

*Complete the following table for **Preventive Action Reports:***

Report Title	Description	Storage Area/Responsible Person (job title)	Length of Storage

Complete the following table for **Company-Level Data:**

Data Type/Title	Description	Storage Area/Responsible Person (job title)	Length of Storage

Complete the following table for **Customer Reference Documents:**

Document Title	Description	Storage Area/Responsible Person (job title)	Length of Storage

Complete the following table for **Purchasing Data:**

Data Type/Title	Description	Storage Area/Responsible Person (job title)	Length of Storage

Complete the following table for **Government Safety and Environmental Reports:**

Report Title	Description	Storage Area/Responsible Person (job title)	Length of Storage

Complete the following table for **Shipping and Packaging Data:**

Data Type/Title	Description	Storage Area/Responsible Person (job title)	Length of Storage

QUALITY RECORDS PROCEDURE TEMPLATE

1. PURPOSE

- *To retain quality records for a specified period.*
- *To identify the storage of quality records that protects them from damage and facilitates retrieval through identification, collection, indexing, and disposition.*
- *To ensure that records are legible, dated (including revision dates), clean, identifiable, and maintained in an orderly manner.*

2. SCOPE

This procedure applies to all records affecting quality and described in the ISO/QS 9000 procedures.

3. RESPONSIBILITIES

<Job title> is responsible for maintaining quality records as defined in this procedure.

<Job title> is responsible for ensuring that copies of documents from superseded parts are available in the new part file.

4. PROCEDURE

4.1 Storage

Quality records are stored <location>. The length of time is determined by <job title> and indicated in the tables below.

Copies of superseded parts documents are placed in the new part file by <job title>.

Quality Records are protected from damage as follows:

- *<Attribute>*

Quality Records are protected from loss as follows:

- *<Attribute>*

Quality Records are legible and readily retrievable. Where computerized, Quality Records follow an established back-up procedure.

<Job title> is responsible for backing up the records.

4.2 Records

The following tables list the Quality Record, where it is stored, and length of storage.

Inspection And Test Records

Title	Description	Storage Area	Length of Storage

Traceability Data

Title	Description	Storage Area	Length of Storage

Contract Review Reports

Title	Description	Storage Area	Length of Storage

Design Review Records

Title	Description	Storage Area	Length of Storage

Training Reports

Title	Description	Storage Area	Length of Storage

Audit Reports

Title	Description	Storage Area	Length of Storage

Nonconforming Product Reports

Title	Description	Storage Area	Length of Storage

Customer-supplied Product Reports

Title	Description	Storage Area	Length of Storage

Calibration Reports

Title	Description	Storage Area	Length of Storage

Process Qualification Reports

Title	Description	Storage Area	Length of Storage

Equipment Verification Reports

Title	Description	Storage Area	Length of Storage

Management Review Reports

Title	Description	Storage Area	Length of Storage

Subcontractor Review Data

Title	Description	Storage Area	Length of Storage

Corrective Action Reports

Title	Description	Storage Area	Length of Storage

Preventive Action Reports

Title	Description	Storage Area	Length of Storage

Company-level Data

Title	Description	Storage Area	Length of Storage

Customer Reference Documents

Title	Description	Storage Area	Length of Storage

Purchasing Data

Title	Description	Storage Area	Length of Storage

Government Safety and Environmental Reports

Title	Description	Storage Area	Length of Storage

Packaging and Shipping Data

Title	Description	Storage Area	Length of Storage

5. RELATED DOCUMENTS

<list work instructions>

Internal Quality Audits

4.17

What is the job title and name of the person responsible for this procedure?

ISO 9000 Standard:

4.17 Internal quality audits

The supplier shall establish and maintain documented procedures for planning and implementing internal quality audits to verify whether quality activities and related results comply with planned arrangements and to determine the effectiveness of the quality system.

Internal quality audits shall be scheduled on the basis of the status and importance of the activity to be audited and shall be carried out by personnel independent of those having direct responsibility for the activity being audited.

The results of the audits shall be recorded (see 4.16) and brought to the attention of the personnel having responsibility in the area audited. The management personnel responsible for the area shall take timely corrective action on deficiencies found during the audit.

Follow-up audit activities shall verify and record the implementation and effectiveness of the corrective action taken (see 4.16).

NOTES

20 The results of internal quality audits form an integral part of the input to management review activities (see 4.1.3).

21 Guidance on quality-system audits is given in ANSI/ASQC Q10011-1-1994, ANSI/ASQC Q10011-2-1994, and ANSI/ASQC Q10011-3-1994.

QS 9000 Interpretations and Supplemental Quality System Requirements

The QS 9000 supplement to ISO 9001, 4.17, "Internal Audit," is Working Environment. This addition requires the supplier to include the working environment as part of the audit criteria. The working environment should be safe, clean, and well-organized.

The Audit Team

Auditors

List the members of the audit team by job title.

Audits must be carried out by personnel not having direct responsibility for the area being audited.

Who (job title) selects auditors for the defined audits?

Are auditors selected from a function not directly involved in the audit? ❐ *yes* ❐ *no*

Do auditors have sufficient seniority in the company to reflect the importance of the audit? ❐ *yes* ❐ *no*

Training

Are auditors trained in auditing techniques? ❐ *yes* ❐ *no*

Management must ensure that verification activities are performed by trained personnel.

If yes, complete the following:

Describe the training (i.e., workshops, on-the-job).

Are records of the training maintained? ❐ *yes* ❐ *no*

If yes, complete the following:

What is the title of the training record?

Where is the record stored?

Who (job title) is responsible for maintaining the training record?

Audit Procedures

You must establish and maintain written audit procedures.

Are auditors provided with written procedures? ❒*yes* ❒ *no*

If yes, complete the following:

Who (job title, department) is responsible for writing the procedure?

Is audit procedure under Document Control? ❒*yes* ❒ *no*

What is the title of the procedure?

Audit Plan

You must schedule audits.

What is the initial frequency of auditing ISO 9000 procedures (i.e., biannually)?

Under what conditions is this frequency adjusted (time interval reduced for new procedures or uncovered deficiencies; interval increased when successful)?

What is the maximum interval between audits?

Who (job title) creates the audit schedule?

Does the audit schedule define:

time frame in which a particular element is to be audited ❒*yes* ❒ *no*
frequency an element is to be audited ❒*yes* ❒ *no*
scope of audits to be performed ❒*yes* ❒ *no*
audit procedures required to perform audits ❒*yes* ❒ *no*

Carry Out Audit

Preparation

Are departments notified of an upcoming audit? ❐*yes* ❐ *no*

If yes, how is the notification given?

Are auditors briefed on procedures and scope prior to the audit? ❐*yes* ❐ *no*

If yes, who (job title) briefs the auditors?

You must have a comprehensive system of audits.

Do auditors study the procedures prior to the audit? ❐*yes* ❐ *no*

Do auditors identify the items of the working environment as part of the audit criteria (i.e., safety, lighting, space) of the audited area?

You must include the working environment as part of the audit criteria.

Record Findings

During the audit, do auditors ensure that procedures are being followed? ❐*yes* ❐ *no*

Do auditors record audit findings? ❐*yes* ❐ *no*

If yes, what is the title of the form to record audit findings?

You must bring audit findings to the attention of the responsible person in the audited area.

Who (job title, review meeting) reviews the audit findings?

Are findings reviewed with the staff responsible for the procedures audited? ❐yes ❐ no

Corrective Action

Implementation

Management must implement timely corrective action.

Upon review of the audit, if an audit finding is legitimate, is the corrective action system followed? ❐ *yes* ❐ *no*

Who (job title) has the final say in cases where the legitimacy of an audit finding cannot be resolved?

Is there a form to complete indicating the corrective action? ❐ *yes* ❐ *no*

If yes, complete the following:

Who (job title) is responsible for the form?

What is the title of the form?

Where is the completed form stored?

Verification

Who (job title) verifies successful implementation of a corrective action?

What is the time period for completing the corrective action?

The implementation and effectiveness of the corrective action must be documented.

Are follow-up audits performed to verify that corrective actions are being maintained? ❐ *yes* ❐ *no*

If yes, complete the following:

What is the time period for completing the follow-up audit?

Who (job title) performs the follow-up audit?

Closing an Audit

What criteria are used to close an audit (i.e., completion of corrective actions, follow-up audit)?

Final Audit Report

Is a final audit report prepared? ❒*yes* ❒ *no*

You must document audit results.

If yes, complete the following:

What is the title of the final audit report?

Who (job title) approves the final audit report prior to distribution?

Who (job title, departments) receives the final audit report?

Final Audit Status Report

Is a report prepared noting that an audit is closed? ❒*yes* ❒ *no*

If yes, complete the following:

What is the title of the report?

Who (job title) is responsible for the report?

Where is the report stored?

Who (job title, departments) receives copies of the report?

Audit History File

Is an audit history file maintained? ❒ *yes* ❒ *no*

If yes, complete the following:

What is included in the audit history file (i.e., reports, written responses, audit findings)?

Who (job title) is responsible for maintaining the audit history file?

What is the title of the audit history file?

Where is the audit history file stored?

Audit Log

Is there a log of audits? ❒ *yes* ❒ *no*

If yes, complete the following:

What is the title of the audit log?

Who (job title) maintains the audit log?

Where is the audit log stored?

What does the log include (i.e., audit start/close date, audit description, applicable procedure number, auditor's name, list of corrective actions, follow-up audit completion date)?

Review and Evaluation

Who (team, departments, job title) reviews audit results?

Are audit results evaluated for the effectiveness of procedures in meeting management's objectives? ❐ *yes* ❐ *no*

Are audit results evaluated to determine whether to modify procedures as a result of new technologies or strategies? ❐ *yes* ❐ *no*

Written Procedures and Related Records

List records or forms, with their corresponding part numbers, used in this procedure:

List related work instructions and procedures, with their corresponding part numbers, that employees use as instructions for activities described above:

INTERNAL QUALITY AUDITS PROCEDURE TEMPLATE

1. PURPOSE

- *To ensure that the procedures in the Quality Manual are being followed.*
- *To determine the effectiveness of the procedures in controlling the quality of <Company>'s products*
- *To identify the need to modify any of the above procedures.*

2. SCOPE

All areas described in the ISO/QS 9000 procedures, policies, and work instructions affecting the quality of work are audited.

3. RESPONSIBILITIES

<Job title>:

- *Initiates the audits and ensures that they are conducted in an efficient manner.*
- *delegates the authority to carry out specific audits in accordance with this procedure.*

<Team>, comprised of <departments> is responsible for forming an internal audit team to audit ISO 9000-compliance.

The cognizant Area Manager responds to audit findings and corrective actions.

4. PROCEDURE

4.1 The Audit Team

The audit team is composed of representatives from the following departments: <departments>.

<Job title> selects auditors for the audits defined.

Auditors are selected from a function not directly involved in the audit. Auditors have sufficient seniority in the company to reflect the importance of the audit.

4.2 Training

Auditors are trained in auditing techniques and attend <training>. <Job title> maintains <training record> in <location> for <length of time>.

Auditors are provided with auditing guidelines for completing the audit by <job title>.

4.3 Audit Plan

Initially, QS 9000 procedures are audited <frequency>. This frequency is adjusted for <activities>. The maximum interval between audits is <interval>.

<Job title> creates the audit schedule. The audit schedule defines:

- *<Attribute>*
- *<Attribute>*
- *<Attribute>*
- *<Attribute>*

4.4 Carry Out Audit

Departments are notified of an upcoming audit by <notification>.

<Job title> briefs auditors on procedures and audit scope prior to the audit, including the working environment criteria. Auditors study the procedures prior to the audit.

During the audit, auditors ensure that procedures are being followed and record their findings onto <name of report>.

<Job title> reviews the audit findings. Audit findings are reviewed with the staff responsible for the procedures audited.

4.5 Corrective Action

Upon review of the audit, if an audit finding is legitimate, the corrective action system is followed.

<Job title> has the final say in cases where the legitimacy of an audit finding cannot be resolved.

<Job title> completes the <name of form> to initiate the corrective action. The <name of form> is stored in <location> for <length of time>.

<Job title> verifies successful implementation of the corrective action. The corrective action must be completed within <time frame>.

Follow-up audits by <job title> within <time frame> verify that the corrective actions are being maintained.

4.6 Closing an Audit

Audits are closed upon <criteria>.

<Job title> prepares a final audit report. <Job title> approves the final audit report prior to distribution. The final audit report is distributed to <names of departments>.

To close the audit, <job title> prepares <name of report>.

<Job title> maintains an audit history file that includes <items>.

<Job title> maintains an audit log stored in <location>. The audit log includes <items>.

4.7 Review and Evaluation

<Team, departments, job title> reviews audit results for <criteria>.

5. RELATED DOCUMENTS

<audit plan>
<audit report>
<audit log>

Management Responsibility
Corrective and Preventive Action
Quality Records

<list work instructions>

Notes:

Training

4.18

What is the job title and name of the person responsible for this procedure?

ISO 9000 Standard:

4.18 Training

The supplier shall establish and maintain documented procedures for identifying training needs and provide for the training of all personnel performing activities affecting quality. Personnel performing specific assigned tasks shall be qualified on the basis of appropriate education, training, and/or experience, as required. Appropriate records of training shall be maintained (see 4.16).

QS 9000 Interpretations and Supplemental Quality System Requirements

The QS 9000 supplements to ISO 9001, 4.18, "Training," is Training as a Strategic Issue. This addition requires the supplier to use training as a method to support the business strategy of the company.

Suggested Procedures

QS 9000:

- Training Effectiveness Evaluation
- Training - Strategic Business Planning

Formal and Informal Training

Training Assessment

Who (job title, i.e., department manager) is responsible for identifying the training needs? **(If there are several employees performing this analysis due to different job titles (salary versus hourly), location or some other factor, please describe the responsibilities of each employee. Also, if different methods are used for determining training needs throughout the organization, please read all questions, and determine if separate procedures are needed for each method. Complete the worksheet as appropriate).**

You must identify training needs.

Are training programs developed to meet employees' needs as identified by the review? ❐*yes* ❐ *no*

Is training provided for employees when excessive nonconformance or quality problems occur? ❐*yes* ❐ *no*

How is training planned for each employee (i.e., based on available courses, employee training plan per position, career path, employee request)?

Orientation

Is orientation provided to new hires? ❐*yes* ❐ *no*

If yes, complete the following:

What does the orientation include (i.e., company orientation, department orientation, product orientation, explanation of quality system)?

Outside Courses

Is successful completion of job-related seminars and college courses recorded in the employee's training record? ❐*yes* ❐ *no*

On-the-job Training

Is on-the-job training provided? ❐yes ❐ no

If yes, complete the following:

Who (job title) provides on-the-job training?

How are the trainers qualified (i.e., experience, trainer certification)?

Safety Training

Is safety training provided? ❐yes ❐ no

If yes, complete the following

What is included in safety training (i.e., first aid, CPR, forklift instruction, back safety)?

Who (job title) provides the safety training?

ESD Training

Is training provided in electro-static discharge awareness? ❐yes ❐ no

Is training provided for hazardous material identification, storage, and handling? ❐yes ❐ no

Personnel

Management Training

Describe the training in quality and QS 9000 that management receives to evaluate the effectiveness of the system (i.e., attend workshops, seminars).

Operators/Assemblers Training

Do assemblers and operators receive (on-the-job) training? ❒ *yes* ❒ *no*

If yes, complete the following:

Who (job title) provides the training?

What is included in the training (i.e., product assembly according to documentation, operation of equipment, quality duties, safety)?

Technical Training

If applicable, complete the following:

What training do technicians receive (on-the-job, seminars, certification by Quality)?

Who (job title) provides the training?

Are training records maintained? ❒*yes* ❒ *no*

What is included in the training (i.e., product repair according to documentation, operation of equipment, quality duties, safety)?

What training do test operators and inspectors receive (on-the-job, seminars, certification by Quality)?

Who (job title) provides the training?

Are training records maintained? ❒*yes* ❒ *no*

What is included in the training (i.e., conducting the test according to documentation, operation of equipment, quality duties, safety)?

What training do calibration specialists receive (on-the-job, seminars, certification by Quality)?

Who (job title) provides the training?

What training do tooling and equipment makers receive (on-the-job, seminars, certification by quality)?

Who (job title) provides the training?

Auditor Training

What training do auditors receive (on-the-job, workshops, certification by Quality)?

Who (job title) provides the training?

What is included in the training?

Job Descriptions

Who (job title) maintains a list of job descriptions?

Personnel must be qualified based on training, education, and experience.

Do job descriptions include the responsibilities and skills required by the employee? ❒*yes* ❒ *no*

Quality Awareness

Does Management ensure that all employees understand the quality policy and how it is implemented and maintained? ❒*yes* ❒ *no*

How is the quality policy communicated to employees (i.e., during training of new hires, during training for new positions, workshops, training certifications, employees learn the quality policies, bonus awards based on the quality of one's work, quality statements posted throughout the facility)?

What is the corrective action for the poor quality of an employee's work (i.e., one-on-one discussions, verbal warnings, written warnings, suspension, termination)?

Records

Training Records

Is the training received by an employee documented? ❐yes ❐ no

You must maintain training records.

If yes, complete the following:

What is the title of the training record?

Who (job title) is responsible for maintaining the training record?

Where is the training record stored?

Who signs the training record?

How often are training records updated?

Training Summary Report

Is there a training summary report? ❐yes ❐ no

If yes, complete the following:

What is the title of the training summary report?

Who (job title) reviews the training summary report?

What is indicated on the report?

Does Management use this report to track the progress of training?

❐yes ❐ no

Training as a Strategic Issue

How is training used to educate employees in future products and processes that are required to maintain company performance in meeting customer expectations?

How is this training implemented throughout the entire organization (i.e., individual training plans)?

How is it determined that training will assist in meeting the goals and objectives of the company and the customer?

How is training effectiveness (i.e., Level 1-4 of effectiveness) measured and evaluated for:

Management

Operators

Skilled Technicians

Support Personnel

Auditors

How are past training results used to determine future training needs?

Written Procedures and Related Records

List records or forms, with their corresponding part numbers, used in this procedure:

List related work instructions and procedures, with their corresponding part numbers, that employees use as instructions for activities described above:

TRAINING PROCEDURE TEMPLATE

1. PURPOSE

- *To ensure that all employees receive adequate training in company procedures, manufacturing skills, service, and safety to meet or exceed customer expectations.*

- *To qualify personnel performing specific tasks on the basis of appropriate education, training, and experience.*

- *To maintain records of training performed.*

- *To provide for a system and instructions and to assign responsibilities for determining training needs, providing the training, and keeping training records.*

- **To provide a process for evaluating the effectiveness of training and deploying training as a strategic issue.**

2. SCOPE

This procedure applies to all training activities at <Company> and affects employees whose work is related to the design, manufacturing, sales, and service of the product as it affects customer relations.

3. RESPONSIBILITIES

<Department> ensures that all new or transfer employees meet the minimum requirements for the position being filled.

<Job title> identifies and ensures that all necessary training courses are received by the staff.

<Job title> provides on-the-job training to operators and assemblers.

4. PROCEDURE

4.1 Formal and Informal Training

<Job title> identifies the training needs.

New hires attend orientation that includes:

- *<Course>*
- *<Course>*
- *<Course>*

Successful completion of job-related seminars and college courses are recorded in the employee's training record.

Employees receive on-the-job training. <Job title> trains employees. Trainers are certified by <attribute>.

4.2 Training Assessment

Training programs are developed to meet employees needs.

When excessive nonconformance or quality problems occur, training is provided for employees.

4.3 Safety Training

Safety training is provided for <items>. <Job title> provides the safety training. Records of attendance are maintained by <job title>.

4.4 ESD Training

Employees attend training in electro-static discharge awareness. Records of attendance are maintained by <job title>.

4.5 Personnel

<Job title> maintains a list of job descriptions. Job descriptions include the responsibilities and skills required by the employee.

4.6 Management Training

Management is trained in quality and ISO 9000 to evaluate the effectiveness of the system by <activity>.

4.7 Operators/Assemblers Training

Assemblers and operators receive training in <activity>. <Job title> provides the training. <Job title> stores the training record in <location> for <length of time>.

4.8 Technical Training

Technicians receive training in <activity>. <Job title> provides the training.<Job title> stores the training record in <location> for <length of time>.

4.9 Auditor Training

Auditors receive training in <activity>. <Job title> provides the training. <Job title> stores the training record in <location> for <length of time>.

4.10 Quality Awareness

Management ensures that all employees understand the quality policy and how it is implemented and maintained through:

- *<Activity>*
- *<Activity>*
- *<Activity>*

4.11 Records

<Training record> is signed by <job title>. <Job title> stores the training record in <location> for <length of time>.

<Job title> generates a <training report>. <Job title> reviews the <training report>. Management uses this report to track the progress of training.

4.12 Training as a Strategic Issue

As part of the business planning process, <job titles> review new and current organizational goals and the relationship of these goals to training.

The effectiveness of previous training to help achieve organizational goals is assessed during the review by:

- **<Activity>**
- **<Activity>**

Training plans are updated as a result of this meeting.

5. RELATED DOCUMENTS

<certificate record>
<training record>
<Business Plan>
<training evaluations>

Management Review
Design Review
Process Control
Inspection and Testing
Inspection, Measuring, and Test Equipment
Internal Quality Audits
Quality Records

<list work instructions>

Notes:

Servicing

4.19

What is the job title and name of the person responsible for this procedure?

ISO 9000 Standard:

4.19 Servicing

Where servicing is a specified requirement, the supplier shall establish and maintain documented procedures for performing, verifying, and reporting that the servicing meets the specified requirements.

QS 9000 Interpretations and Supplemental Quality System Requirements

The QS 9000 supplement to ISO 9001, 4.19, "Servicing," dictates that a procedure must be implemented that ensures that service issues (i.e., customer complaints and feedback) are communicated to the manufacturing, engineering, and design activities.

Suggested Procedures:

- *International Service*
- *Domestic Service*
- *Customer Service*

QS 9000:

- Corrective Action - Service

General

How is maintenance of the product stipulated (i.e., maintenance agreement policy, government contract)?

Specification

Do you provide warranties on your product? ❒*yes* ❒ *no*

Do you sell extended warranties for the product? ❒*yes* ❒ *no*

Do you sell maintenance agreements? ❒*yes* ❒ *no*

If yes, complete the following:

Who (job title, department) generates the agreement with the customer?

Where is the maintenance agreement filed?

Does the agreement specify:

items to be maintained ❒*yes* ❒ *no*
period of time ❒*yes* ❒ *no*
other

Manuals and Written Procedures

Is there a service manual or documented procedures? ❒yes ❒ no

If yes, complete the following:

Is the documentation a controlled document? ❒yes ❒ no

What does the documentation describe (i.e., installation, operation, parts)?

Who (job title) uses the service documentation (i.e., field service technician)?

Training

Customer Support

Do you provide training for customer support? ❒yes ❒ no

What is customer support trained to do?

Are training records maintained? ❒yes ❒ no

If yes, complete the following:

Where are the records stored?

How long are the records kept?

Service Technicians

Do you provide a training program for service technicians in your employment? ❐*yes* ❐ *no*

If yes, complete the following:

What is the technician qualified to do (i.e., adjustments, repairs, alignment, calibration, verification)?

At the time of training, what is issued to technicians (i.e., tools, test equipment, service manuals)?

Is additional training provided with new product introduction? ❐*yes* ❐ *no*

What is the format of the training (i.e., hands-on, individualized by Product Manager, on-the-job with experienced technicians)?

Training Records

Are training records maintained? ❐*yes* ❐ *no*

If yes, complete the following:

Where are the training records stored?

How long are the training records kept?

Service Representatives

Who is eligible for service training, if applicable (i.e., military customers, foreign sales force, factory representatives)?

How are service representatives from international locations trained (i.e., seminars provided by your company, trained by field service during installation)?

Tools

Are tools and equipment used to service product in the field appropriate for the task? ❐*yes* ❐ *no*

Do tools and equipment pass the same calibration standards as tools used in the manufacturing facility? ❐*yes* ❐ *no*

If yes, complete the following:

Who (job title) verifies the calibration of tools?

Are calibration stickers applied to the tools? ❐*yes* ❐ *no*

Are tool records kept on file? ❐*yes* ❐ *no*

If yes, where are the records stored?

Dissemination

Incoming Customer Calls

Who (job title) handles incoming customer calls?

Is the call assigned for service? ❐ *yes* ❐ *no*

Is a log of the call maintained? ❐ *yes* ❐ *no*

If yes, complete the following:

What is the name of the log?

What is identified on the log?

Customer Support

Is an attempt made to solve the problem through telephone or fax? ❐*yes* ❐ *no*

If yes, complete the following:

Who (job title) communicates with the customer?

What records of the customer call are maintained?

What does the record indicate (i.e., description of problem, attempted resolution, date, authorizing person, on-site contact person, assigned number)?

What happens if the product or part is under warranty (i.e., replacement sent to customer, representative replaces part)?

What happens if the product or part is not under warranty (replacement sent, written agreement with customer authorizing costs for repair)?

What is the usual time frame for service?

Records

Service Record

What is the name of the record that documents the service activity?

What is included in the record?

Where is the completed record kept?

How long is the completed record kept?

What is included in the record (i.e., name, type, date, serial number, purpose of call, action taken, customer signature, P.O. number, parts used, labor time, expenses, customer comments)?

Who (job title) approves the report?

What other service records are maintained (i.e., Bill of Materials for replacement parts)?

Outside Service Record

Is service performed by factory representatives documented? ❐ *yes* ❐ *no*

If yes, complete the following:

What is the name of the record?

What is included (i.e., customer name, address, and phone number, equipment model and serial number, warranty status, customer-reported complaint, service technician's diagnosis, action taken/parts used to repair, amount of time spent to repair, customer's signature confirming service performed, date of service call)?

Follow-up

Contact

Upon completion of repair, is the customer contacted to determine satisfaction? ❐ *yes* ❐ *no*

If yes, complete the following:

Is a card sent to the customer? ❐ *yes* ❐ *no*

If yes, what is the name of the card? ❐ *yes* ❐ *no*

Does the customer provide comments for the service report, if applicable? ❐ *yes* ❐ *no*

Customer Feedback

Is there a procedure for communicating customer feedback to engineering? ❒yes ❒ no

Is there a procedure for communicating customer feedback to manufacturing? ❒yes ❒ no

Is there a procedure for communicating customer feedback to design? ❒yes ❒ no

If yes, to the above questions, complete the following:

Who (job title) prepares this procedure?

Who (job title) maintains this procedure?

Is a form used to record customer feedback? ❒yes ❒ no

What is the name of the form?

Who (job title) reviews customer feedback?

What contact is made with the customer (i.e., letter of response, replacement of warranted parts)?

You must implement a procedure that communicates customer feedback to engineering, design, and manufacturing.

Product Verification

Failure Data

Is statistical analysis kept on failure data? ❒*yes* ❒ *no*

If yes, who (job title) reviews the data?

Product Monitoring

Is performance of the product monitored throughout its life cycle? ❒*yes* ❒ *no*

If yes, complete the following:

What is monitored?

final calibration records compared to calibrations made during service call ❐*yes* ❐ *no*
continuing operation ❐*yes* ❐ *no*
customer satisfaction, including safety and reliability ❐*yes* ❐ *no*
parts whose life cycle expires, replaced as part of service contract ❐*yes* ❐ *no*

Credits

Are credits issued, when applicable? ❐*yes* ❐ *no*

If yes, complete the following:

Which situation(s) constitute a credit?

over-shipment or incorrect shipment ❐*yes* ❐ *no*
product didn't perform to specification ❐*yes* ❐ *no*
customer returned product prior to using it ❐*yes* ❐ *no*
other

Who is responsible for issuing the credit?

Written Procedures and Related Records

List records or forms, with their corresponding part numbers, used in this procedure:

List related work instructions and procedures, with their corresponding part numbers, that employees use as instructions for activities described above:

SERVICING PROCEDURE TEMPLATE

1. PURPOSE

- *To control the quality of service provided by the technical service representatives and field technicians.*
- *To control the service and re-manufacturing of returned devices.*
- *To control the sale and service of maintenance agreements.*
- *To control the disposition of material returned to <Company> customers and dealers.*
- *To document material movement and disposition through the return process.*
- **To establish and implement a procedure for communicating customer feedback to engineering, design, and manufacturing.**

2. SCOPE

This procedure applies to service provided by field service technicians, and service representatives, as well as service provided through customer service calls for the domestic and international market.

This procedure covers return authorizations for <name of product>.

3. RESPONSIBILITIES

<Job title> ensures calibration of test and measuring equipment.

<Job title> services product and completes reports of their activity.

<Job title> issues and updates technical service manuals for products manufactured and sold by <Company>.

<Job title> trains service employees and qualifies third-party service organizations to perform repairs.

<Job title> documents the receipt of goods into the repair department and assigns a preliminary disposition to returned material and services units, subassemblies, and accessories dispositioned to repair.

<Job title> maintains warranty and extended warranty records.

<Job title> reviews customer feedback and communicates the feedback to engineering, design, and manufacturing.

4. PROCEDURE

4.1 Specification

The <type of warranty/contract> stipulates the product maintenance. Customers may purchase extended warranties and maintenance agreements.

<Job title, department> generates the agreement with the customer. The maintenance agreement is filed <location> and specifies:

- *<Item>*
- *<Item>*
- *<Item>*

4.2 Manuals and Written Procedures

<Name of documentation>, under document control, describes <items> and is used by <job title> in the servicing of product.

4.3 Training

Customer support representatives receive training to <activity>. <Job title> maintains training records in <location> for <length of time>.

Service technicians employed by <Company> are trained as follows:

- *<Training>*
- *<Training>*
- *<Training>*

The technician is qualified to <activity>. At the time of training, the technician receives <items>. Additional training is provided with new product introduction. <Job title> maintains training records in <location> for <length of time>.

4.4 Tools

Tools and equipment used to service product in the field are appropriate for the task and pass the same calibration standards as tools used in the manufacturing facility.

<Job title> verifies the calibration of tools. Calibration stickers identify the calibration date and identification. Records of tools are kept on file and stored <location>.

4.5 Dissemination

<Job title> handles incoming customer calls and assigns the call for service. A log of the call is maintained in the <name of log> and includes <information>.

<Job title> communicates through telephone or fax to attempt to solve the problem. <Job title> maintains a record of the customer call indicating <information>. **This information is communicated to design, engineering, and manufacturing by:**

- <Activity>
- <Activity>
- <Activity>

If the product or part is under warranty, <activity> occurs.

If the product or part is not under warranty, <activity> occurs.

The usual time frame for service is <length of time>.

4.6 Records

<Job title> documents the service activity on <name of report> and includes <items>. The completed <name of report> is stored <location> for <length of time>. <Job title> approves the report. Other service records, such as <documents>, are stored with the report.

Service performed by factory representatives is documented on <name of report> and includes <items>.

4.7 Follow-up

Upon completion of a repair, <type of contact> is made with the customer to determine satisfaction.

The customer provides comments for the service report, if applicable.

<Job title> reviews customer complaints and contacts the customer as follows: <contact>.

4.8 Product Verification

<Job title> keeps statistical analysis on failure data. <Job title> reviews the failure data.

Performance of the product is monitored throughout its life cycle as follows:

- *<Review activity>*
- *<Review activity>*
- *<Review activity>*

4.9 Credits

Where applicable, <job title> issues a credit for:

- *<Event>*
- *<Event>*
- *<Event>*

5. RELATED DOCUMENTS

<telephone call log>
<scheduling log>
<service report
<service records>
<material inspection and receiving report>
<service critique card>
<returned goods evaluation form>
<service manuals>

Contract Review
Document and Data Control
Process Control
Inspection, Measuring, and Test Equipment
Control of Nonconforming Product
Training
Statistical Techniques
Corrective and Preventive Action

<list work instructions>

Notes:

Statistical Techniques

4.20

What is the job title and name of the person responsible for this procedure?

ISO 9000 Standard:

4.20 Statistical techniques

4.20.1 Identification of need

The supplier shall identify the need for statistical techniques required for establishing, controlling, and verifying process capability and product characteristics.

4.20.2 Procedures

The supplier shall establish and maintain documented procedures to implement and control the application of the statistical techniques identified in 4.20.1.

QS 9000 Interpretations and Supplemental Quality System Requirements

The QS 9000 supplements to ISO 9001, 4.20, "Statistical Techniques," are:

- Selection of Statistical Tools
- Knowledge of Basic Statistical Concepts

These additions require the supplier to prepare control plans for each process requiring statistical analysis. Suppliers are encouraged to ensure that employees know basic statistical concepts. Suppliers should use the Advanced Product Quality Planning and Control Plan Reference Manual and the Statistical Process Control Reference Manual.

Suggested Procedures

QS 9000:

- Selection and Analysis of Statistical Tools
- Preparation of Control Plans
- Advanced Product Quality Planning

Applications

Complete the following table with the inspections or tests from which you analyze data:

Test or Inspection Name	Responsible Person	Collection Method
(i.e., receiving inspection)	*(i.e., inspector)*	*(i.e., visual)*
(i.e., in-process test)	*(i.e., operator)*	*(i.e., test equipment)*
(i.e., final system test)	*(i.e., Manufacturing Engineer)*	*(i.e., software tools interfacing with system operation)*

Process Analysis

Types of Analysis

Complete the following table with the types of analysis, test, and record-keeping:

Type of Analysis	Test/Inspection Data	Records
(i.e., first pass yield, burn-in failure rates, Pareto, histogram)		

Data Collection

For each analysis in the table above, complete the following:

Who (job title) collects the data?

How often is the data collected?

How is the data collected (i.e., sampling reports, barcode scanner, inspection reports)?

Test	Responsible Person	Frequency	Collection Method

Review

Who (job title/departments) reviews and analyzes the data?

How is the data disseminated (i.e., chart displayed in facility, presented at meetings)?

Sampling Plan

Is a sampling plan used at inspection? ❐*yes* ❐ *no*

If yes, what is the sampling plan or where is it specified?

Data Review

Who (i.e., any employee, SPC group member, Test Engineer) can collect data?

Are meetings held to review the data? ❐*yes* ❐ *no*

If yes, complete the following:

What is the name of the meeting?

What is the frequency of the meetings?

Who (job titles, departments) attend the meeting?

What records are kept (i.e., minutes, reports)?

Are causes of problems and corrective actions reviewed at the meeting? ❐*yes* ❐ *no*

Market Analysis

Is marketing data collected? ❐ *yes* ❐ *no*

If yes, complete the following:

Who (job title) collects marketing data?

What types of data are collected (i.e., competitor's market share, customer feedback)?

From where are the data collected (i.e., survey, studies, customer complaints)?

Who (job title) reviews the marketing data?

Training

What type of training is provided (i.e., on-the-job, courses)?

Who (job titles) receives statistical training?

Do these courses include the basics of statistical process control (i.e., variation, stability, capability, overadjustment)? ❐ yes ❐ no

What training records are maintained?

Who (job title) maintains the training records?

Documented Procedures and Control Plans

Are there documented procedures for collecting and analyzing statistical data? ❒yes ❒ no

If yes, complete the following:

Is the Advanced Product Quality Planning and Control Plan Reference Manual used in selecting the tool for collecting the data? ❒yes ❒ no

You must use control plans.

Is the Fundamentals of Statistical Process Control Reference Manual used in collecting and analyzing statistical data? ❒yes ❒ no

Is the documentation controlled? ❒yes ❒ no

What does the documentation describe (i.e., data collection, data analysis, data review)?

Who (job title) is responsible for creating the documentation?

Written Procedures and Related Records

List records or forms, with their corresponding part numbers, used in this procedure:

List related work instructions and procedures, with their corresponding part numbers, that employees use as instructions for activities described above:

STATISTICAL TECHNIQUES PROCEDURE TEMPLATE

1. PURPOSE

- *To ensure consistent quality and appropriate process control through the use of statistical techniques.*
- *To collect, analyze, and interpret data relating to product and process performance.*

2. SCOPE

This procedure applies to data gathered from process, shipment, and inspections and from marketing analysis.

3. RESPONSIBILITIES

<Job title> collects data from in-process tests.

<Job title> samples the quality of received product.

<Job title> identifies all critical process control points and statistical methods to be employed.

<Job title> collects test results from barcode readers.

<Job title> prepares control plans.

4. PROCEDURE

4.1 Applications

The following table lists the inspections or tests from which data are analyzed:

Test or Inspection Name	Responsible Person	Collection Method
(i.e., receiving inspection)	*(i.e., inspector)*	*(i.e., visual)*
(i.e., in-process)	*(i.e., operator)*	*(i.e., test equipment)*
(i.e., final system test)	*(i.e., Manufacturing Engineer)*	*(i.e., software)*

4.2 Process Analysis

The following table lists the types of analysis, test, and record-keeping:

Types of Analysis	Tests /Inspection Data	Records
(i.e., first pass yield, burn-in failure rates, Pareto, histogram)	<tests>	<records>

The following table lists the person responsible for data collection, frequency, and method:

Test	Responsible Person	Frequency	Collection Method

<Job title, departments> reviews and analyzes the data. The data are disseminated by <method>.

4.3 Data Review

<Job title> collects data. <Frequency>, at <name of meeting> attended by <job titles>, the data and causes of problems and corrective actions are reviewed.

4.4 Market Analysis

<Job title> collects marketing data through <analysis>. <Job title> reviews the marketing data.

4.5 Training

Personnel performing data collection or analysis are qualified by:

- Activity
- Activity

4.6 Documented Procedures and Control Plans

<Job title> documents the procedures for collecting and analyzing statistical data. These procedures are maintained under the Document Control system.

Control plans are prepared by <job title> for processes requiring statistical analysis.

5. RELATED DOCUMENTS

<control plans>
<control charts>
Contract Review
Process Control
Inspection and Testing
Servicing
Management Responsibility

<list work instructions>

Production Part Approval Process

II.I

What is the job title and name of the person responsible for this procedure?

This QS 9000 element requires the supplier to complete the Production Part Approval Process to the levels required by the customer. The Production Part Approval Process (PPAP) Reference Manual describes the documentation and process that must be completed. Production Part Approval packages must be submitted to the customer for approval prior to the first shipment of production parts.

Chrysler, General Motors, and Ford have specific requirements for production part approval. These requirements are in Appendix B of the Production Part Approval Process Reference Manual. Basically, these requirements describe specific forms or processes for production approval or changes. Suppliers need to incorporate each of the customer's requirements into the PPAP procedure.

Suggested Procedures

QS 9000:

- Material Tests
- Performance Tests
- Product Layout and Verification
- Statistical Techniques
- Capability Studies
- Control Plans
- Control Charts
- Failure Mode Effects Analysis

General

Parts for production approval must be manufactured using production processes, materials, and equipment.

Has the customer designated the supplier as a "self-certifying supplier?"
❐yes ❐ no

Is production part approval conducted for automotive parts? ❐yes ❐ no

If yes, complete the following:

Who (job title, i.e., Lead Engineer, Program Manager, Quality Manager) is responsible for coordinating the PPAP?

Production part approvals must be maintained for the length of time that the part is in production or service, plus one calendar year.

Where (location) are PPAP packages maintained?

Is there a written procedure for PPAP? ❐yes ❐ no

How long are PPAP records maintained?

Submission Requirements

Are PPAP packages submitted according to the customer-required submission level?

❐yes ❐ no

Who (job title) is responsible for obtaining the submission level requirement from the customer?

Are the customer-designated submission level requirements documented?

❐yes ❐ no

PPAP Submittals

Complete the following. Is PPAP conducted when:

new part	❐yes	❐ no
new product	❐yes	❐ no
change to product	❐yes	❐ no
change to process	❐yes	❐ no
change in material	❐yes	❐ no
change in design specifications	❐yes	❐ no
use of previously approved material	❐yes	❐ no
new tooling	❐yes	❐ no
modified tooling	❐yes	❐ no
refurbished tooling	❐yes	❐ no
tooling is relocated	❐yes	❐ no
equipment is moved to another part of the plant	❐yes	❐ no
equipment is moved to another facility	❐yes	❐ no
product is manufactured in another facility	❐yes	❐ no
product is manufactured on a line that has not been approved	❐yes	❐ no
change in subcontractor of material	❐yes	❐ no
change in subcontractor parts	❐yes	❐ no
change in subcontractor process	❐yes	❐ no
change of subcontractor	❐yes	❐ no
tooling is used that has been inactive for more than 12 months	❐yes	❐ no
customer requests	❐yes	❐ no

If yes, complete the following:

Where are the customer submission level requirements maintained (i.e., PPAP file)?

If the customer does not supply the submission level, what level is used (i.e., Level 3)?

You must obtain prior customer approval for computer-generated replicas of the forms in the PPAP Reference Manual.

Is a copy of the Part/Submission Warrant Form (CFG-1001) used for PPAP submittal?
❒yes ❒ no

Is there a supplier-created part submission warrant form? ❒yes ❒ no

If yes, complete the following:

Is the form an exact replica of the form in the PPAP Reference Manual?
❒yes ❒ no

Is customer approval of the form obtained prior to using it for submission?
❒yes ❒ no

Is a copy of the Appearance Approval Report Form (CFG-1002) used for PPAP submittal?
❒yes ❒ no

Is there a supplier-created Appearance Approval Report Form (CFG-1002)?
❒yes ❒ no

If yes, complete the following:

Is the form an exact replica of the form in the PPAP Reference Manual?
❒yes ❒ no

Is customer approval of the form obtained prior to using it for submission?
❒yes ❒ no

Is a copy of the Dimensional Results Form (CFG-1003) used for PPAP submittal?
❒yes ❒ no

Is there a supplier-created Dimensional Results Form? ❒yes ❒ no

If yes, complete the following:

Is the form an exact replica of the form in the PPAP Reference Manual?
❒yes ❒ no

Is customer approval of the form obtained prior to using it for submission?
❒yes ❒ no

Is a copy of the Material Test Results Form (CFG-1004) used for PPAP submittal?
❐yes ❐ no

Is there a supplier-created Material Test Results Form? ❐yes ❐ no

If yes, complete the following:

Is the form an exact replica of the form in the PPAP Reference Manual?
❐yes ❐ no

Is customer approval of the form obtained prior to using it for submission?
❐yes ❐ no

Part Approval Requirements

Complete the following table indicating the person (job title) responsible for providing the information required for part approval:

PPAP Requirement	Responsible Employee
Submission warrant	
Appearance Approval Report	
Master sample	
Customer supplied design records/drawings	
Supplier design records/drawings	
Authorized engineering change documents	
Dimensional results	
Checking aids	
Material test reports	
Performance test reports	
Durability test reports	
Process flow diagram	
Process FMEA	
Design FMEA	
Control Plans for significant characteristics	
Process capability studies	
Measurement system variation studies	
Engineering approval on customer drawings/ specifications	

You must mark all supporting information and retain documents according to submission level requirements.

Auxiliary Drawings and Sketches

Are all supporting drawings and sketches marked with the following information:

part number	❒yes	❒ no
change level	❒yes	❒ no
drawing date	❒yes	❒ no
supplier name	❒yes	❒ no

Are the supporting documents submitted and retained according to the submission level requirements? ❒yes ❒ no

Is a tracing provided to the customer approval activity when an optical comparator is required for inspection? ❒yes ❒ no

Part-specific Inspection or Test Devices

You must provide part-specific test/inspection equipment if requested by the customer.

Is there product-specific test equipment? ❐yes ❐ no

If yes, complete the following:

Is a certificate provided that assures all aspects of the gage agrees with part dimensional requirements? ❐yes ❐ no

Is that certification provided to the customer with the submission? ❐yes ❐ no

Are revisions of the product tracked to assure that engineering changes are incorporated into the test/inspection device? ❐yes ❐ no

If yes, complete the following:

Is review of the revision incorporated into the test/inspection devices documentation? ❐yes ❐ no

What is the name of the form used to document the change review?

Where is the document maintained?

Who (job title) is responsible for maintaining the change record?

Who (job title) reviews the engineering change?

Who (job title) authorizes changes to the inspection/test devices?

Are gage studies (i.e., Gage R&R, accuracy, linearity) conducted according to customer requirements? ❐yes ❐ no

If no, list and describe the measurement system variation studies that are performed:

Customer-identified Special Characteristics

Are customer-specific markings used for special characteristics? ❒yes ❒ no

If no, what markings are used to identify customer-designated special characteristics?

Preliminary Process Capability Studies

You must perform Ppk studies for special characteristics.

Are Preliminary Process Capability (Ppk) studies performed for all special characteristics that can be measured with variable data? ❒yes ❒ no

Are measurements system analysis studies conducted on special characteristic items? ❒yes ❒ no

Are control plans developed to provide instructions on the measurements that must be performed for determining Ppk? ❒yes ❒ no

Are control charts, specifically X-bar and R charts, used to record inspection /test results? ❒yes ❒ no

If yes, complete the following:

Is a minimum of 25 subgroups used? ❒yes ❒ no

Is a minimum of 100 individual readings made? ❒yes ❒ no

Is the control chart examined for stability? ❒yes ❒ no

If yes, is corrective action taken for unstable conditions? ❒yes ❒ no

Where (i.e., reaction plan, corrective action request) is the corrective action documented?

Who (job title) is responsible for completing the corrective action form?

Where is the corrective action form maintained?

Is the customer contacted if the process is not stable? ❒yes ❒ no

What actions are taken if the calculated $Ppk > 1.67$?

What actions are taken if $1.33 \leq Ppk \leq 1.67$?

What actions are taken if $Ppk < 1.33$?

Corrective action must be taken on unstable process.

What actions are taken when a process is unstable (i.e., identify special causes, 100% inspection)?

If a process is considered unstable, is the corrective action documented?
❒yes ❒ no

If yes, complete the following:

What is the name of the form?

Who (job title) prepares the form?

Who (job title) monitors the status of the corrective action?

Who (job title) approves the corrective action?

Is the control plan revised to reflect the interim corrective actions (i.e., 100% inspection)?
❒yes ❒ no

If yes, complete the following:

Is the control plan reviewed and approved by the customer? ❒yes ❒ no

Who (job title) updates the control plan?

Who (job title) submits the control plan for customer approval?

How is the customer's approval documented and tracked (i.e., control plan signed off by customer)?

Appearance Approval Requirements (AAR)

You must complete an AAR for each item or series of items.

Are there customer-designated appearance items? ❒yes ❒ no

If no, go to the next section.

If yes, complete the following:

Is a separate AAR prepared for each part or series of parts? ❒yes ❒ no

Who (job title) is responsible for preparing the AAR?

Is the AAR submitted to the customer according to the customer-specific requirements?
❒yes ❒ no

Are representative production parts submitted to the customer-designated location?
❒yes ❒ no

Is the AAR submitted with the PPAP package? ❒yes ❒ no

If yes, complete the following:

Does the completed AAR include the customer's signature? ❐yes ❐ no

Does the completed AAR include part disposition instructions? ❐yes ❐ no

Who (job title) assures that the disposition instructions for the AAR item are followed?

Dimensional Evaluations

You must perform dimensional inspections.

Are dimensional inspections performed on all parts that have dimensional requirements? ❐yes ❐ no

Who (job title) performs the dimensional inspections?

Subcontracted dimensional inspection results must be submitted in the subcontractor's normal format.

Are dimensional inspections subcontracted? ❐yes ❐ no

If yes, complete the following:

Are the inspection reports in subcontractor format or on subcontractor letterhead? ❐yes ❐ no

Is the subcontractor's name indicated on the report? ❐yes ❐ no

Are actual dimensional results for each characteristic recorded on the report? ❐yes ❐ no

What is the name of the form that is used to record the results (i.e., Dimensional Results Form, checked drawing)?

Are the dates that the inspections were conducted indicated on the report? ❐yes ❐ no

Is the date of the design record indicated on the report? ❐yes ❐ no

Is the change level of the approved part/drawing on the report? ❐yes ❐ no

Does the report contain a list of any approved engineering changes that are not incorporated into the test item? ❐yes ❐ no

Are hard copy results of math data submitted with the PPAP submittal? ❐yes ❐ no

If yes, does the hard copy indicate where the measurements were taken? ❐yes ❐ no

Multiple Tools, Molds, Dies and Patterns

You must submit complete dimensional analyses of parts produced from different molds, tools, dies, or patterns.

Will product be produced by more than one die, mold, tool, or pattern? ❐yes ❐ no

If yes, is a dimensional analysis conducted on parts from each of the tools, molds, patterns, and dies? ❐yes ❐ no

Master Samples

You must maintain complete records and a master sample for each submittal.

Is one of the items identified as the master sample? ❐yes ❐ no

What (i.e., tag, label) is used to designate the master sample?

Does the master sample include the customer approval date? ❐yes ❐ no

Is the master sample retained according to the submission level requirements? ❐yes ❐ no

Is corrective action taken if the part or product does not meet the dimensional requirement? ❐yes ❐ no

Is the corrective action documented? ❐yes ❐ no

If yes, complete the following:

What is used to document the corrective action (i.e., Nonconforming Material Report, corrective action request)?

Who (job title) prepares the report?

Is the customer notified if the part or product does not meet any of the dimensional requirements? ❒yes ❒ no

Who (job title) is responsible for notifying the customer of the nonconformities?

Is a form used to notify the customer? ❒yes ❒ no

If yes, complete the following:

What is the name of the form (i.e., Request for Customer Approval, Supplier Request for Engineering Approval)?

Who (job title) prepares the form?

Where is the form maintained?

Part Weight

You must determine the weight of the part.

Is the weight of the part provided to the customer? ❒yes ❒ no

Is the weight of the part expressed in kilograms to three decimal places (e.g., 0.000) ❒yes ❒ no

Is the weight calculated by obtaining the average of ten randomly selected parts? ❒yes ❒ no

Does the part weigh less than 0.100 kilograms? ❒yes ❒ no

If yes, is the part weight obtained by weighing ten parts together and calculating the average weight?

Material tests must be conducted.

Material Tests

Are chemical, physical, and/or metallurgical requirements specified for the product?
❒yes ❒ no

If yes, are material tests performed to determine conformance to the material specification of the part or product? ❒yes ❒ no

Are any of the material tests subcontracted? ❒yes ❒ no

If yes, complete the following:

Are the test reports in subcontractor format or on subcontractor letterhead?
❒yes ❒ no

Is the subcontractor's name indicated on the report? ❒yes ❒ no

Are actual test results for each characteristic recorded on the report? ❒yes ❒ no

What is the name of the form that is used to record the results (i.e., Material Test Results Form)?

Is the number of items tested on the report? ❒yes ❒ no

Is the date that the test(s) was performed on the report? ❒yes ❒ no

Is the drawing and engineering change level of the part indicated on the report?
❒yes ❒ no

Is the date that the test was performed on the report? ❒yes ❒ no

Is the material supplier's name on the report? ❒yes ❒ no

Is the material supplier a customer-approved subcontractor? ❒yes ❒ no

If yes, is the code number for the material indicated on the report? ❒yes ❒ no

Is the change level of approved part/drawing on the report? ❒yes ❒ no

Does the report contain a list of any approved engineering changes that were not incorporated into the test item? ❒yes ❒ no

Is corrective action taken if the part or product does not meet the material test requirements?
❐yes ❐ no

Is the corrective action documented? ❐yes ❐ no

If yes, complete the following:

What is used to document the corrective action (i.e., Nonconforming Material Report, corrective action request)?

Who (job title) prepares the report?

Is the customer notified if the part or product does not meet any of the material requirements?
❐yes ❐ no

Who (job title) is responsible for notifying the customer of the nonconformance?

Is a form used to notify the customer? ❐yes ❐ no

If yes, complete the following:

What is the name of the form (i.e., Request for Customer Approval, Supplier Request for Engineering Approval)?

Who (job title) prepares the form?

Where is the form maintained?

Performance Tests

Performance tests must be conducted.

Are performance requirements specified for the product? ❒yes ❒ no

If yes, are performance tests conducted to determine conformance to the performance specifications and control plan of the part or product? ❒yes ❒ no

Are any of the performance tests subcontracted? ❒yes ❒ no

If yes, complete the following:

Are the sources for the performances qualified (i.e., approved laboratories) to perform the test? ❒yes ❒ no

Are the test reports in subcontractor format or on subcontractor letterhead? ❒yes ❒ no

Is the subcontractor's name indicated on the report? ❒yes ❒ no

Are actual test results for each characteristic recorded on the report? ❒yes ❒ no

What is the name of the form that is used to record the results (i.e., Performance Test Results Form)?

Is the number of items tested on the report? ❒yes ❒ no

Is the date that the test(s) was performed on the report? ❒yes ❒ no

Is the drawing and engineering change level of the part indicated on the report? ❒yes ❒ no

Does the report contain a list of any approved engineering changes that were not incorporated into the test item? ❒yes ❒ no

Is corrective action taken if the part or product does not meet the performance test requirements? ❒yes ❒ no

Is the corrective action documented? ❒yes ❒ no

If yes, complete the following:

What is used to document the corrective action (i.e., Nonconforming Material Report, corrective action request)?

Who (job title) prepares the report?

Is the customer notified if the part or product does not meet any of the performance requirements? ❒yes ❒ no

Who (job title) is responsible for notifying the customer of the nonconformance?

Is a form used to notify the customer? ❒yes ❒ no

If yes, complete the following:

What is the name of the form (i.e., Request for Customer Approval, Supplier Request for Engineering Approval)?

Who (job title) prepares the form?

Where is the form maintained?

Part Submission Warrant

Who (job title) completes the Part Submission Warrant?

You must complete the Part Submission Warrant after successful completion of the requirements.

Who (job title, i.e., Quality Manager) is responsible for assuring the Warrant is complete according to customer-specific requirements?

Who (job title) is responsible for signing the Warrant indicating that the report is complete, submitted to the proper submission level, the proper supporting information is attached, and that all customer requirements are met?

Engineering Changes

Are PPAP packages submitted when there is an engineering change? ❐yes ❐ no

Who (job title) is responsible for contacting the customer's part approval authority to obtain guidance for the information required in the submittal?

For engineering changes, complete the following:

Who (job title) completes the Part Submission Warrant?

Who (job title, i.e., Quality Manager) is responsible for assuring the Warrant is complete according to customer-specific requirements?

Who (job title) is responsible for signing the Warrant indicating that the report is complete, is submitted to the proper submission level, the proper supporting information is attached, and that all customer requirements are met?

PPAP Records

You must maintain records of all submittals.

Are records maintained to indicate the results of all tests and inspection required as part of the submission? ❐yes ❐ no

Do the records maintained for each submittal include:

spc results	❐yes	❐ no
appearance approval	❐yes	❐ no
inspection results /design record	❐yes	❐ no
lab test results	❐yes	❐ no
material test results	❐yes	❐ no
ppk studies	❐yes	❐ no
msa results	❐yes	❐ no
process flow diagrams	❐yes	❐ no
PFMEAs	❐yes	❐ no
DFMEAs	❐yes	❐ no
control plans	❐yes	❐ no
subcontractor warrants	❐yes	❐ no

performance evaluations ❐yes ❐ no
master sample ❐yes ❐ no

How long are PPAP records maintained (i.e., active production and service plus one calendar year)?

Who (job title) maintains the PPAP records?

How long are master samples maintained?

Who (job title) maintains the master sample?

Where (location) are the master samples maintained?

Part Submission Status

Interim Approval

If interim approval is granted by the customer, is the root cause of the nonconformities identified? ❐yes ❐ no

Is an interim approval action report prepared? ❐yes ❐ no

If yes, complete the following:

Who (job title) prepares the interim approval action report?

Who (job title) approves the interim approval action report?

Who (job title) tracks the status of the corrective action?

Rejected PPAP Submittals

If the PPAP is rejected, is the cause for rejection researched to identify root cause?

❐yes ❐ no

Is the PPAP resubmitted once the product meets all customer requirements?

❐yes ❐ no

You must follow customer-specific requirements for PPAP

Customer-specific Requirements for PPAP

Have you reviewed customer-specific requirements for PPAP submittal?

❐yes ❐ no

If no, review the PPAP Reference Manual and incorporate the appropriate customer-specific requirements.

PRODUCTION PART APPROVAL PROCESS PROCEDURE TEMPLATE

1. PURPOSE

- To assure that automotive production and service product meets the customer's requirements prior to production.
- To assign responsibility for the preparation, conduct, and reporting of product qualification tests and inspections.

2. SCOPE

This procedure applies to product designed, manufactured, assembled, and shipped to automotive production and service locations.

3. RESPONSIBILITIES

<Job title> prepares Part Submission Warrants and assures that all production part approval requirements are met.

<Job title> approves the Production Part Approval submittal.

<Job title> maintains records and samples of production part approval.

<Job title> prepares corrective action reports for nonconforming production part approval results.

4. PROCEDURE

4.1 Production Part Approval Process (PPAP) Submittals

The Production Part Approval Process is conducted on product provided to <customer names> for production or service. <Job title> coordinates the necessary tests, inspections, and reports for the PPAP submittal.

PPAP submittals are prepared according to the requirements of the most current edition of the Production Part Approval Process Reference Manual and this procedure.

The PPAP is conducted or updated under any of the following conditions:

- New product
- Change to product
- Change to process
- Change in material
- Change in design specifications
- Use of previously approved material
- New tooling
- Modified tooling
- Refurbished tooling
- Tooling is relocated
- Tooling is used that has been inactive for more than 12 months
- Equipment is moved to another part of the plant
- Equipment is moved to another facility
- Product is manufactured in another facility
- Product is manufactured on a line that has not been approved
- Change of subcontractor
- Change in subcontractor of material
- Change in subcontractor parts'
- Change in subcontractor process
- Customer requests
- New part

4.2 Selection of Parts for PPAP Tests and Validations

Parts used for production part approval are a representative sample run from the same processes, tooling, and equipment that will produce the production parts.

The quantity of parts obtained for production approval will be from a <list type of operation> with target quantity of 300 unless another quantity is specified by the customer.

When parts are produced using different tooling, molds, patterns, or dies, representative parts from each of these will be selected by <job title> for measurement and testing as part of the production approval process.

4.3 PPAP Submittals

Reports are prepared to demonstrate conformance to the customer's requirement. The following table identifies the personnel responsible for assuring that test and inspections are performed, results recorded, and that the customer's requirements are met.

PPAP Document	Responsible Employee
Submission warrant	<Job title>
Appearance Approval Report	<Job title>
Sample part	<Job title>
Customer-supplied design records/drawings	<Job title>
Supplier design records/drawings	<Job title>
Authorized engineering change documents	<Job title>
Dimensional results	<Job title>
Checking aids	<Job title>
Material test reports	<Job title>
Performance test reports	<Job title>
Durability test reports	<Job title>
Process flow diagram	<Job title>
Process FMEA	<Job title>
Design FMEA (if applicable)	<Job title>
Control plans for significant characteristics	<Job title>
Process capability studies	<Job title>
Measurement system variation studies	<Job title>
Engineering approval on customer drawing/ specification	<Job title>

4.4 PPAP Report Identification

Dimensional evaluations, material tests, and performance test records are marked with:

- **<Marking>**
- **<Marking>**
- **<Marking>**

The results of dimensional evaluations are recorded on <name of form> and indicate the actual evaluation results for each characteristic evaluated.

The results of material tests are recorded on <name of form> and indicate the actual test results for each test conducted.

The results of performance tests are recorded on<name of form> and indicate the actual test results.

If dimensional evaluations and/or material or performance tests are subcontracted, the subcontractor's report is marked by <marking> and indicates:

- **<Item>**
- **<Item>**
- **<Item>**

4.5 Subcontracted Items

Part or product subcontractors are notified of PPAP requirements by <method>.

Subcontractors must comply with the PPAP requirements stated in the <form>.

<Job title> is responsible for assuring that subcontractors provide information and results according to the customer's requirements.

4.6 Corrective Action

<Job title> is responsible for obtaining corrective action on product, tests, and processes that do not meet customer requirements.

If interim approval is granted by the customer, <job title> prepares an interim approval action plan for customer approval and assures that the action plan is followed. <Job title> petitions the customer if an extension to the interim approval is needed. Product will only be shipped to the customer that is covered by the interim approval allowances.

Corrective action for rejected or nonconforming PPAP submittals will be done according to Corrective Action Procedure, xxx.

If a PPAP submittal is rejected, corrected product and documentation will be resubmitted to the customer for approval. Production quantities will not be shipped on rejected PPAP submittals until approval of the PPAP is obtained from the customer.

4.7 <Customer Name> Requirements

For PPAP submittals to <Customer>, <job title> assures that all customer- specific product and documentation requirements are met prior to submittal. These requirements include:

<Requirement>

<Requirement>

<Requirement>

For changes to <Customer> products or processes that have obtained PPAP approval, <job title> coordinates these changes according to the customer-specific requirements.

Appendix B of the PPAP Reference Manual is used to identify customer- specific requirements.

4.8 PPAP Records

All PPAP submittals are submitted and retained according to the submission level requirements specified by the customer. A record of the customer's submission level designation is retained in the <location>.

<Job title> prepares and signs the Part Submission Warrant for <Company>. This signature indicates that the part meets all customer production part approval requirements.

The record of the PPAP submittal and all supporting documentation is maintained by <job title> for the length of time that the part is active for either production or service plus one calendar year. The record of PPAP approval is maintained by<job title> in <location> for a minimum of <length of time>.

The master sample is retained as part of the submittal records and according to the submission level requirements specified in the <u>PPAP Reference Manual</u>.

<Job title> maintains the master sample for the length of time that the part is active for either production or service plus one calendar year, or until a new master sample is made to meet customer requirements. The master sample is stored in <location>, is marked by <method>, and includes the customer's approval date.

5. RELATED DOCUMENTS

<part submission warrant form>
<appearance approval report>
<dimensional evaluation results>
<material test results>
<performance test results>
<customer specific reports>
<capability studies>
<process flow diagrams>
<control plans>
<control charts>
<process failure mode and effects analysis>
<design failure mode and effects analysis>
<design aids>

Corrective and Preventive Action
Design Control
Statistical Techniques

<list work instructions>

Continuous Improvement

II.2

What is the job title and name of the person responsible for this procedure?

The QS 9000 requires suppliers to establish a continuous improvement program and to deploy a continuous improvement philosophy throughout the organization. For processes that are stable, capable, and produce items that are important to the customer, suppliers must develop specific improvement plans.

Suppliers must also identify opportunities for continuous improvement throughout the organization. Improvement projects must be implemented to enhance overall quality and productivity.

The supplier must demonstrate knowledge of advanced process improvement techniques such as benchmarking, cost of quality, and value analysis. These techniques must be used appropriately.

Suggested Procedures:

- Business Planning
- Benchmarking
- Value Analysis
- Cost of Quality
- Mistake-proofing

General

Does the organization have a formal continuous improvement philosophy?

❐yes ❐ no

You must have a fully deployed continuous improvement philosophy.

Does the continuous improvement philosophy include:

quality?	❐yes	❐ no
service?	❐yes	❐ no
delivery?	❐yes	❐ no
price?	❐yes	❐ no
business planning?	❐yes	❐ no
administrative functions?	❐yes	❐ no

Is there a name for the continuous improvement program? ❐yes ❐ no

If yes, what is the name of the program?

You must identify and implement quality and productivity improvement projects.

Identification of Quality and Productivity Improvement Opportunities

Are quality and productivity improvement projects identified? ❐yes ❐ no

Who (job title) can identify an opportunity for improvement?

What mechanism is used to identify the improvement opportunity (i.e., employee suggestion forms, idea forms, work group meetings)?

Who (job title) approves quality and productivity improvement ideas?

Are teams developed to work on quality and productivity improvements? ❐yes ❐ no

Who (job title, any employee) can participate on quality and productivity improvement teams?

Who (job title) maintains a list of quality and productivity improvement ideas?

Where is the list maintained?

How often (frequency) is the status of the quality and productivity ideas reviewed (i.e., monthly staff meetings, quarterly quality and productivity steering team meetings, annual management review)?

Employees must be knowledgeable in continuous improvement techniques.

Process Improvement Plans

Are there specific process improvement action plans for processes that are important to the customer once the processes are stable and capable? ❒yes ❒ no

You must have specific continuous improvement action plans.

If yes, complete the following:

What are the names or form numbers for these plans?

Who (job title) prepares the plans?

Who (job title) tracks the progress of the plans?

Who (job title, team name) reviews the status of the plans?

Where are the improvement plans maintained?

Continuous Improvement Techniques

Continuous Improvement Tools

Are appropriate employees knowledgeable of the following methods:

capability indices ❒yes ❒ no
- control charts ❒yes ❒ no
- cumulative sum charting ❒yes ❒ no
- design of experiments ❒yes ❒ no
- evolutionary operation of processes ❒yes ❒ no
- theory of constraints ❒yes ❒ no
- overall equipment effectiveness ❒yes ❒ no
- cost of quality ❒yes ❒ no
- parts per million analysis ❒yes ❒ no
- value analysis ❒yes ❒ no
- problem-solving ❒yes ❒ no
- benchmarking ❒yes ❒ no
- analysis of motion/ergonomics ❒yes ❒ no
- mistake-proofing ❒yes ❒ no

Are these methods used in quality and productivity improvements? ❐yes ❐ no

Are employees who have knowledge of the above methodologies identified? ❐yes ❐ no

Is training provided to employees in these methods? ❐yes ❐ no

Continuous Improvement Teams

How (method) are employees identified that have knowledge of these methods(i.e., training database, employee file, position summary)?

Are these employees available to participate on quality and productivity teams? ❐yes ❐ no

If these employees are not available to participate, is there an alternate method (i.e., facilitation, participation by a subject matter expert) that allows these methodologies to be used in quality and productivity improvements?

Are the results of quality and productivity improvements incorporated into the analysis of company-level data? ❐yes ❐ no

Are the results of quality and productivity improvements supportive of the overall business planning process? ❐yes ❐ no

CONTINUOUS IMPROVEMENT PROCEDURE TEMPLATE

1. PURPOSE

- To establish and implement a company approach to continuous improvement.
- To assure the deployment of a continuous improvement philosophy throughout the organization.

2. SCOPE

This procedure applies to the processes and products of <Company>.

3. RESPONSIBILITIES

<Any employee> identifies a continuous improvement opportunity.

<Job title> maintains the <name of continuous improvement opportunity log>.

<Job title> maintains the <continuous improvement plans for stable, capable processes log>.

<Management> provides the necessary resources to continuous improvement teams and reviews continuous improvement projects. Management incorporates continuous improvement ideas and projects into business planning and the analysis of company-level data.

<Any employee, customer, or supplier> may participate in a continuous improvement project if identified by <job title or team name>.

<Job title, Team name> reviews, approves, disapproves, or sends to a team, continuous improvement opportunities.

<Job title, Team name> reports the status of continuous improvement projects to employees.

4. PROCEDURE

4.1 Process Improvement Plans

For processes that are stable and capable, <job title, team name> prepares an action plan that identifies continuous improvement opportunities for stable and capable processes. The <job title, team name> continues to improve the process and report results to Management by <method and frequency>. These plans are maintained by <job title> in <location>.

4.2 Identification of Continuous Improvement Opportunities

Continuous improvement opportunities in quality and productivity can be identified by any employee. The use of customer surveys, customer concerns, FMEA, the business planning process, and management reviews are encouraged when identifying projects.

An employee prepares and submits <name of continuous improvement form> to <job title, team name> for review. <Job title> logs the continuous improvement idea into the <name of log>.

Continuous improvement opportunities and the results of continuous improvement projects are reported to employees through:

- <Activity>
- <Activity>
- <Activity>

Management reviews the results of continuous improvement projects <frequency> at <event, method>. The projects and opportunities are also reviewed as part of the business planning process.

4.3 Continuous Improvement Tools

Employees participating in continuous improvement projects are encouraged to use prevention- and prediction-based tools. These tools are:

- Capability indices
- Control charts
- Cumulative sum charting
- Design of experiments
- Evolutionary operation of processes
- Theory of constraints
- Overall equipment effectiveness
- Cost of quality
- Parts per million analysis
- Value analysis
- Problem-solving
- Benchmarking
- Analysis of motion/ergonomics
- Mistake-proofing

A list of employees trained in the above methodologies is maintained by <department, job title>. Any employee with the appropriate training may be designated as a team member or may serve as a subject matter expert or facilitator to a continuous improvement team. Employee <training plans, records> are modified to facilitate proper knowledge of these techniques.

4.4 Continuous Improvement Teams

<Job title, Team name> approves, disapproves, or submits the continuous improvement idea to a team for further research. The team is responsible for developing a project plan for incorporating the opportunity and reporting the status to Management on a <frequency basis> by <method>.

Continuous improvement project teams are cross-functional. Members are identified by <method>. Team members are allowed to work on projects during their workday and these activities are considered part of their job duties. Management provides the resources to support these teams.

5. RELATED DOCUMENTS

<continuous improvement idea form>
<continuous improvement status log>
<process action log>
<training plans>
<training records>

Management Responsibility
Training
Process Control
Corrective and Preventative Action

<list work instructions>

Manufacturing Capabilities

II.3

What is the job title and name of the person responsible for this procedure?

This section of the QS 9000 requires suppliers to address the effectiveness of facility, equipment, and process planning. Suppliers are required to use cross-functional teams during the Advanced Product Quality Planning Process for developing equipment, facilities, and process plans. The effectiveness of the existing manufacturing systems must also be evaluated. Suppliers must also use mistake-proofing techniques during manufacturing planning.

Tooling must be managed and planned. Resources must be available for the maintenance and use of tools and equipment. Customer-owned tools and equipment must be permanently marked. The supplier must establish and implement an effective tooling management program.

Suggested procedures

- Facility and Equipment Effectiveness
- Mistake-proofing
- Tooling Maintenance
- Tooling Set-up
- Tooling Storage

Facility Planning

Is there a current system for manufacturing planning? ❐yes ❐ no

If yes, complete the following:

Does the system work in conjunction with the Advanced Product Quality Planning Process?
❐yes ❐ no

Are cross-functional teams used for facility/manufacturing planning? ❐yes ❐ no

Cross-functional teams must conduct facility planning

Who (job titles) comprise the facility planning team?

How often does the team meet?

Does the team consider the following during facility planning:

material travel and handling	❐yes	❐ no
synchronous material flow	❐yes	❐ no
value-added floor space	❐yes	❐ no

What other topics are considered during facility planning?

Facility Effectiveness

Is there a team that evaluates the effectiveness of current operations? ❐yes ❐ no

You must evaluate the effectiveness of existing operations.

If yes, complete the following:

Does the team consider the following topics when evaluating effectiveness:

overall work plan	❐yes	❐ no
automation	❐yes	❐ no
ergonomics	❐yes	❐ no
human factors	❐yes	❐ no
operator and line balance	❐yes	❐ no
storage and buffer inventory levels	❐yes	❐ no
value-added labor content	❐yes	❐ no

How often is facility effectiveness formally assessed (annually, prior to a new process being considered)?

Is facility effectiveness incorporated into continuous improvement projects?
❒yes ❒ no

Is facility effectiveness incorporated into the business planning process?
❒yes ❒ no

Who (job title) is responsible for monitoring facility effectiveness?

How is facility effectiveness evaluated (i.e., measurables established for quarterly review, measurables established for analysis and use of company-level data)?

Mistake-proofing must be considered when nonconformances are identified during process planning.

Is mistake-proofing used by the facility? ❒yes ❒ no

When is mistake-proofing used (i.e., whenever potential product nonconformities are identified during the planning process)?

Who (job title, team name) is responsible for assuring that mistake-proofing is used during process planning?

Tooling Management

You must establish and implement a tooling management system.

Is there a system for tooling management? ❒yes ❒ no

If yes, complete the following:

Who (job title) is responsible for tooling management?

Does the tooling management system consider maintenance and repair facilities?
❒yes ❒ no

Is there a master list of tools, molds, and dies used by the organization?
❒yes ❒ no

Who (job title) maintains the list?

What is the name of the list?

Where is the list stored?

Does the tooling management system include replacement and repair capabilities for critical tooling? ❐yes ❐ no

Are personnel trained in the use of tools? ❐yes ❐ no

Are personnel trained in tool, mold and die design, fabrication, and maintenance? ❐yes ❐ no

Where (department) are the records maintained for personnel involved in tooling?

Who (job title) is responsible for maintaining tooling training records?

Does the tooling management system include storage, preservation, and recovery of tooling? ❐yes ❐ no

Is there a tool change program for perishable tools? ❐yes ❐ no

Who (job title) monitors tool wear?

What criteria are used in establishing the change intervals for tooling?

Who (job title) is responsible for maintaining the following?

tooling

molds

dies

Are there procedures for tooling set-up? ❐yes ❐ no

Are the procedures in accordance with 4.9.5 of the QS 9000? ❐yes ❐ no

Who (job title) prepares the procedures?

Where are the procedures maintained?

Who (job title) maintains the set-up procedures?

Who (job title) validates the set-up procedures?

Are any of the tooling management functions subcontracted? ❒yes ❒ no

You must track any subcontracted tooling activities.

If yes, complete the following:

Is there a tracking system to assure proper tool maintenance and repair? ❒yes ❒ no

Is there a tracking system for tool storage, preservation, and recovery? ❒yes ❒ no

Is there a tracking system for set-up? ❒yes ❒ no

Is there a tracking procedure to ensure that perishable tooling is properly changed? ❒yes ❒ no

Tool Design, Fabrication, and Repair

Are adequate resources available for tool and gage design (i.e., qualified personnel, precision measurement tools)? ❐yes ❐ no

You must have adequate resources to support tool design.

Are adequate resources available for the fabrication of tools? ❐yes ❐ no

Who (job title, team name) supplies the resources?

Are adequate resources available for dimensional inspections? ❐yes ❐ no

Who (job title) repairs:

tools

dies

molds

Customer-owned Tooling and Equipment

Is customer-owned tooling and equipment permanently marked? ❐yes ❐ no

Are the markings visible on each item? ❐yes ❐ no

Customer-owned tooling and equipment must be visible marked.

If yes, complete the following:

Who (job title) is responsible for marking the items?

How is the marking applied (i.e., non-removable label, etching, stamp)?

Is there a log of customer-supplied tooling and equipment?

If yes, complete the following:

Who (job title) maintains the log?

What is the name of the log?

Where (location) is the log maintained?

MANUFACTURING CAPABILITIES PROCEDURE TEMPLATE

1. PURPOSE

- **To establish a system that utilizes cross-functional teams for manufacturing planning, including processes and facilities.**
- **To establish a tooling management system.**
- **To assure effective implementation of prevention-based planning in determining manufacturing capabilities.**
- **To establish a method for evaluating the effectiveness of existing manufacturing operations.**

2. SCOPE

This procedure applies to manufacturing processes at <Company>.

3. RESPONSIBILITIES

<Any employee> may participate on a manufacturing planning team.

<Job title, Team name> establishes cross-functional manufacturing teams.

<Job title, Team name> prepares manufacturing effectiveness measures.

<Job title> presents corrective action resolutions or status to <Management>.

<Job title> is responsible for the tooling management system.

<Job title> maintains the <tool/die/mold log>.

<Job title> repairs tools/dies/molds.

<Job title> performs tooling set-up.

<Job title> maintains tooling set-up procedures and records.

<Job title> monitors perishable tool wear.

<Job title> assigns, applies, and assures the marking of customer-owned tooling.

<Management> reviews manufacturing effectiveness measures, assigns corrective action team members, and provides the necessary resources for tooling management.

4. PROCEDURE

4.1 Facility Planning

<Job title, Team name> forms cross-functional teams for the manufacturing planning of new processes/facilities or changes to existing processes/facilities. Minimally, the team consists of:

- **<Job title>**
- **<Job title>**
- **<Job title>**
- **<Job title>**
- **<Job title>**

The <team name> reviews and assesses the capability of the processes to efficiently produce conforming product. These activities include, but are not limited to:

- **<Activity>**
- **<Activity>**
- **<Activity>**
- **<Activity>**

The <team name> uses mistake-proofing of the process or product to prevent the manufacture or assembly of nonconforming product. Potential nonconformances identified by FMEA, capability studies, or service reports are addressed by mistake-proofing techniques.

4.2 Facility Effectiveness

The effectiveness of current manufacturing processes is measured by:

- **<Method>**
- **<Method>**
- **<Method>**
- **<Method>**

<Job title, Team name> is responsible for preparing the effectiveness measures and reporting the status to Management at <event>.

If processes exhibit a decline in effectiveness, a cross-functional continuous improvement team is assigned by Management to investigate the cause of the decline and to correct the problem. <Job title> presents the status of the problem resolution to Management.

4.3 Tooling Management

<Job title> is responsible for the tooling management system. The tooling management system assures that the proper tools, gages, molds, and dies are available for manufacturing.

Tools are maintained by <job title>. Repairs to tools are conducted by <job title, approved subcontractor>. Personnel performing tool repair, modification, or fabrication are qualified by <source or type of qualification>. Records of the qualification are maintained by <job title>. <Name of log> identifies the tools in <Company>'s tooling system and is maintained by <job title> in <location>.

Dies are maintained by <job title>. Repairs to dies are conducted by <job title, approved subcontractor>. Personnel performing die repair, modification, or fabrication are qualified by <source or type of qualification>. Records of the qualification are maintained by <job title>. <Name of log> identifies the dies in <Company>'s tooling system.

Molds are maintained by <job title>. Repairs to molds are conducted by <job title, approved subcontractor>. Personnel performing mold repair, modification, or manufacture are qualified by<source or type of qualification>. Records of the qualification are maintained by <job title>. <Name of log> identifies the molds in <Company >'s tooling system.

Tooling set-up is performed by <job title, department> according to documented procedures. Verification of the set-up is conducted by <list verification criteria from 4.9.5>.

<Job title> monitors process and tool wear to establish the tool change program for perishable tools. Tool wear is recorded in <name of log>. The log is maintained by <job title> in <location>. Indications of excessive tool wear are investigated by <job title, team name>. The cause of the excessive tool wear is determined and corrective action is taken.

Procedures are maintained by <job title> for the storage and recovery of tools. The recovery and storage procedures are stored in <location>.

4.4 Tool Design, Fabrication, and Repair

Management provides the resources necessary for tool design fabrication. These resources include design, fabrication, repair, and inspection activities.

Subcontracted tool design, fabrication, and repair is tracked by <job title>. <Job title> ensures that the supplier:

- <Activity>
- <Activity>
- <Activity>

4.5 Customer-owned Tooling and Equipment

Customer-owned tooling and equipment is indicated by <marking>. <Job title> applies the marking when the equipment is logged into the <name of log>.

The marking is readily apparent and is on customer-owned tooling and equipment. Customer-owned tooling and equipment is logged into <name of log>.

5. RELATED DOCUMENTS

<record of manufacturing effectiveness measures>
<tools, molds, dies log>
<tool wear log>
<customer owned tooling log>
<tool set-up procedures>
<training records>

Advanced Product Quality Planning and Control Plan
Tool Storage And Recovery Procedures
Continuous Improvement
Corrective and Preventive Action
Quality Records
Training

<list work instructions>

Customer-Specific Requirements

III

What is the job title and name of the person responsible for this procedure?

This section of the QS 9000 contains the requirements that are specific to Chrysler, General Motors, and Ford. Each of the automobile manufacturers has supplemented the first two sections of the QS 9000 with information contained in this section.

To assist in developing procedures to support the customer-specific requirements, the following tables identify the customer paragraph and the appropriate elements of the QS 9000 that should address the requirement. Suppliers may write procedures specific to the customer requirements or incorporate these requirements into the procedures supporting the requirement. The worksheets for each section have already asked questions that are applicable to each of the customer-specific requirements. These tables will help you review your existing procedures to make certain that the customer requirements are addressed.

QS 9000 Customer-Specific Procedure Cross-Reference

Supplier/Item	4.1	4.2	4.3	4.4	4.5	4.6	4.7	4.8	4.9	4.10	4.11	4.12	4.13	4.14	4.15	4.16	4.17	4.18	4.19	4.20	II.1	II.2	II.3
GENERAL MOTORS																							
Key Characteristics		X		X					X		X									X	X		
Supplier Submission of Material																				X			
Problem Reporting														X								X	
Supplier Submission-Match																				X			
Component Verification								X															
Continuous Improvement																					X		
Run at Rate									X													X	
Evaluation Supplier Test										X	X									X			
Early Production Containment										X			X	X						X			
Traceability Identifier								X															
Specs for Bar Codes ECV/VCVS															X					X			
Procedures for Prototype Material		X		X					X														
Packaging/ID for Production															X					X			
Shipping/Parts ID															X					X			
Shipping/Delivery Performance															X								
Cust. Approval - Control Plans		X		X					X											X			
UPC Labeling															X								
Layout Inspection										X										X			

Supplier/Item	4.1	4.2	4.3	4.4	4.5	4.6	4.7	4.8	4.9	4.10	4.11	4.12	4.13	4.14	4.15	4.16	4.17	4.18	4.19	4.20	II.1	II.2	II.3
FORD																							
Control Plans/FMEAs				X	X				X		X									X	X		
Shipping Container Label															X								
Equipment Stand. Parts				X		X																	
Critical Characteristics		X		X					X												X	X	
Set-up Verification										X													
Material Analysis - Heat Treat		X						X	X														
Material Analysis - Non-Heat Treat								X															
Lot Traceability								X															
Heat Treating									X														X
Process Changes				X	X			X	X												X	X	X
Supplier Modification		X			X				X	X		X											
ES Test Requirements				X					X			X	X	X							X		
System Design Specs				X																			
Ongoing Process Monitoring									X											X		X	
Prototype Part Quality		X							X												X		
QOS	X																					X	
Qualification/Acceptance				X	X				X												X		

Supplier/Item	4.1	4.2	4.3	4.4	4.5	4.6	4.7	4.8	4.9	4.10	4.11	4.12	4.13	4.14	4.15	4.16	4.17	4.18	4.19	4.20	II.1	II.2	II.3
CHRYSLER																							
Significant Characteristics		X		X					X		X									X	X		
Annual Layout										X													
Internal Quality Audit																	X						
Design Val./Product Ver.				X															X				
Corrective Action														X							X	X	X
Packaging, Shipping, Labeling															X								
Process Sign-off									X												X		X
Lot Acceptance Sampling										X		X	X	X						X			

Installing the Software

Appendix A

What's in This Section?

The disk included with this book is available in formats for Microsoft Word for Windows, WordPerfect for Windows, and Microsoft Word for Macintosh. Read this section to learn:

- how to install the software
- the directory structure and files to use
- how to open files

If Using Word or WordPerfect for Windows ...

This section instructs you in how to install the disk, shows the directory structure, and explains how to open a file.

Installing the Disk

To install the Microsoft Word or WordPerfect for Windows version on a personal computer:

1. Power on your computer.
2. Open Windows.
3. Insert the disk into the A drive.
4. Open the **File Manager**.
5. Click on the **A drive** icon.

   The window displays the file structure of the diskette, as shown in the following illustration.

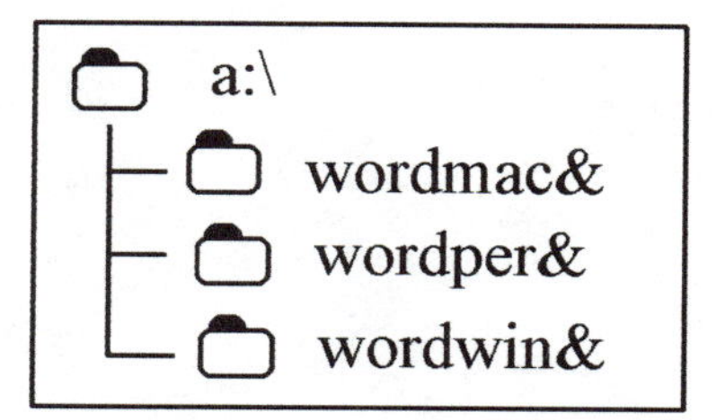

6. Click on **wordwin&** or **wordper&.**

   The highlight bar appears on the selection.

7. From the File menu, select **Copy**.

The window displays the Copy dialog box, shown in the following illustration.

Copy

Current Directory: A:\WORDWIN& or WORDPER&

From: A:\WORDWIN& or \WORDPER&

To:

OK

Cancel

Help

8. Move the I-beam cursor to the To box and type:

   C:\winword\QS 9000 or **C:\wpwin60\QS 9000**

9. Click the **OK** button.

   The system copies the files from the disk to the subdirectory QS 9000. This subdirectory is located in the winword or wpwin60 directory.

10. Close the **File Manager.**

This completes the installation. Remove the disk from the A drive and store it safely.

Understanding the File Structure

Your directory structure looks like this:

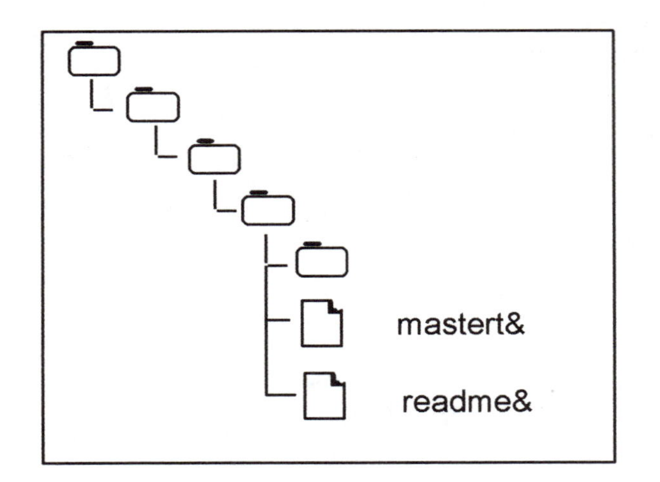

Table A-1 identifies the files in the subdirectory named "procedu&" (shortened from "procedures.")

File Name	QS Element
4-10i.doc	Inspection and Testing
4-11t. doc	Inspection, Test, and Measurement Equipment
4-12t. doc	Inspection and Test Status
4-13n. doc	Control of Nonconforming Product
4-14c. doc	Corrective and Preventive Action
4-15h. doc	Handling, Storage, Packaging, Preservation, and Delivery
4-16q. doc	Quality Records
4-17i. doc	Internal Quality Audits
4-18t. doc	Training
4-19s. doc	Servicing
4-1mg. doc	Management Responsibility
4-20s. doc	Statistical Techniques
4-2qu. doc	Quality System
4-3co. doc	Contract Review
4-4de. doc	Design Control
4-5do. doc	Document and Data Control
4-6pu. doc	Purchasing
4-7cu. doc	Customer-Supplied Product
4-8pr. doc	Product Identification and Traceability
4-9pr. doc	Process Control
contimp. doc	Continuous Improvement
mfgcap. doc	Manufacturing Capabilities
ppap. doc	Production Part Approval Process

Table A-1. File Names and Corresponding QS 9000 Elements

Opening Files

To open a file:

Note: If your system does not show any files, select the all files option in the dialog box.

1. Double-click on the Microsoft Word or WordPerfect icon.

2. From the File menu, select **Open**.

3. Double-click on the directory **QS 9000**.

4. Double-click on the directory **wordwin&** or **wordper&**.

   The files master.doc and readme.doc appear in the list box. To access these files, double-click on the desired file.

5. Double-click on the directory **procedu**&.

   A list of files, as shown in Table A-1, appears.

6. Double-click on the desired file.

If Using Word for the Macintosh ...

This section instructs you in how to install the disk, shows the directory structure, and explains how to open a file.

You must have software installed on your Macintosh that allows you to read DOS-formatted disks.

Installing the Disk

To install the Microsoft Word for Macintosh version:

1. Power on your computer.
2. Insert the disk into the disk drive.

   The disk icon, titled QS 9000, appears on your desktop.

3. Create a new folder to hold the templates.
   a) From the Finder File menu, select **New Folder**.
   b) Type a folder name and press Return. For example, you may want to create a folder on your hard drive entitled QS 9000.
4. Double-click the disk icon on your desktop.

   A window opens, displaying the folders WordMac, WordPerfect, and Word Windows.

5. Drag the WordMac folder to the new folder you just created.

   The system copies the procedure and master templates to the new folder.

This completes the installation. Remove the disk from the drive and store it safely.

Understanding the File Structure

Inside the folder Word Mac are two files and another folder.

Your directory structure looks like this:

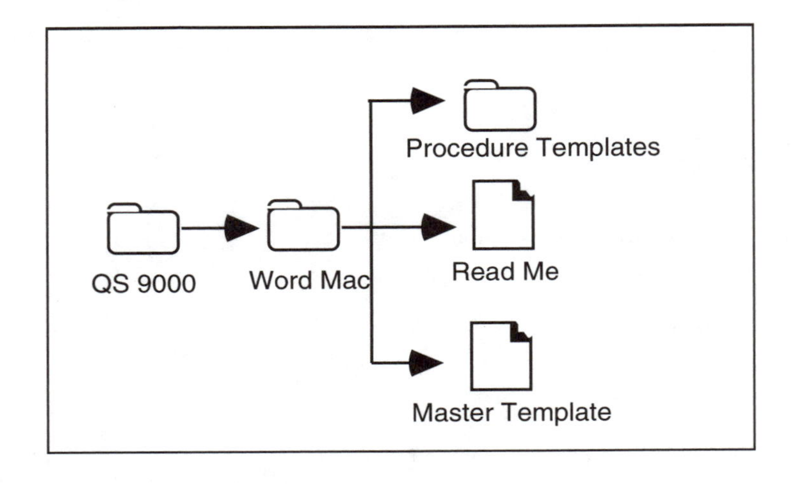

Table A-2 identifies the files in the folder named Procedure Templates.

File Name	QS Element
4-10i.mcw	Inspection and Testing
4-11t. mcw	Inspection, Test, and Measurement Equipment
4-12t. mcw	Inspection and Test Status
4-13n. mcw	Control of Nonconforming Product
4-14c. mcw	Corrective and Preventive Action
4-15h. mcw	Handling, Storage, Packaging, Preservation, and Delivery
4-16q. mcw	Quality Records
4-17i. mcw	Internal Quality Audits
4-18t. mcw	Training
4-19s. mcw	Servicing
4-1mg. mcw	Management Responsibility
4-20s. mcw	Statistical Techniques
4-2qu. mcw	Quality System
4-3co. mcw	Contract Review
4-4de. mcw	Design Control
4-5do. mcw	Document and Data Control
4-6pu. mcw	Purchasing
4-7cu. mcw	Customer-Supplied Product
4-8pr. mcw	Product Identification and Traceability
4-9pr. mcw	Process Control
contimp.mcw	Continuous Improvement
mfgcap.mcw	Manufacturing Capabilities
ppap.mcw	Production Part Approval Process

Table A-2. File Names and Corresponding QS 9000 Elements

Opening Files

To open a file:

1. Open Microsoft Word.
2. From the File menu, select **Open**.
3. From the Select a Document pop-down menu, select the folder that holds the folder QS 9000.
4. Double-click on **QS 9000**. The WordMac folder appears on the list.
5. Double-click on **WordMac**.

   The files Master Template and Read Me, and the folder Procedure Templates, appear in the list box.

6. To open a procedure template:

   a) Double-click on the folder **Procedure Templates**.

      A list of files, as shown in Table A-2 appears.

   b) Double-click on the desired file.

Procedure Status Form

Appendix B

What's in This Section?

This section provides a form for you to complete as you track your progress. Use this form to enter the procedure title, originator, and draft/review status of your procedures. For more information, refer to Section 2, "Planning Your QS 9000 System and Documentation."

Procedure Status

Title	Originator	Distribute Worksheet	Review Worksheets	1st Draft Complete	Reviewers' Comments	2nd Draft Complete	Sign-off

QS 9000 Accreditation Bodies

Appendix C

What's in This Section?

The following is a list of QS 9000 accreditation bodies as of March 22, 1996. This listing was obtained from the *IASG Sanctioned QS 9000 Interpretations, March 22, 1996.* A revised list is published with each IASG update. Each accreditation body maintains a list of qualified registrars.

Registrar Accreditation Board (RAB)
Paul Fortlage
611 East Wisconsin Ave
Milwaukee, WI 53201-3005
phone: 414-272-8575
fax: 414-765-8661

United Kingdom Accreditation Services (UKAS)
Robin Bullock-Webster
13 Palace Street
London, UK SW1 E 5HS
phone: 44-071-233-7111
fax: 44-071-233-5115

Dutch Council for Certification (RvA)
Peter Goosen
Stationsweg 13F
NL - 3972 KA Driebergen
The Netherlands
phone: 31-3438-12604
fax: 31-3438-18554

Joint Accreditation System of Australia and New Zealand
Steve Keeling
P.O. Box 164 Civic Square ACT Australia
51 Allara Street Canberra ACT Australia
phone (international): 616-276-1999
fax: (international): 616-276-2041

SWEDAC
Lars Ettarp
P.O. Box 878, SE 501 15
Boras, Sweden
phone: 46-33-17-77-45
fax: 46-33-10-13-92

TGA - Tragergemeinschaft Fur Akkreditierung GMBH
Dr. Thomas Facklam
/buro: Stresemannallee 13
60596 Frankfurt am Main, Germany
phone: 49-69-630-09111
fax: 49-69-630-09144

Swiss Accreditation Service (SAS)
J.-P. Jaunin
CH-3084 Wabern
Lindenweg 50, Switzerland
phone: 41-31-963-31-11
fax: 41-31-963-32-10

Entidad Nacional de Acreditacion (ENAC)
Fernando Maiz De La Torre
Serrano, 240.28016
Madrid, Spain
phone: 91-457-32-89
fax: 91-458-62-80

Accreditamento Organismi Certificazione (SINCERT)
Enrico Martinotti
Via Battistotti Sassi, 11
21033 Milano
phone: 02-719202-719664
fax: 02-719055

Standards Council of Canada (SCC)
Joan Brough-Kerrebyn
1200-45 O'Conner
Ottawa, Ontario, Canada K1P 6N7
phone: 613-238-3222
fax: 613-995-4564

Centre for Metrology and Accreditation (FINAS)
Tuulikki Hattula
P.O. Box 239 (Lonnrotinkatu 37),
FIN-00181 Helsinki
phone: 358-0-61-671
fax: 358-0-61-67341

Japan Accreditation Board (JAB)
Takashi Otsubo
Akasaka Royal Bldg Annex
6-18 Akasaka 7 Chome, Minato-ku
Tokyo 107, Japan
phone: 81-3-5561-0375
fax: 81-3-5561-0376

Justervesenet-Norweigian Metrology and Accreditation Service
Leif Halbro
P.O. Box 6832 St. Olavs Plass
N- 1030 OSLO, Norway
phone: 47-22-20-02-26
fax: 47-22-20-77-72

Republik Osterreich (BMwA)
Dipl.-Ing. Gunter P. Friers
A-1031 Wien, Landstr, Hauptstr. 55-57
DVR 37 257, Vienna, Austria
phone: 43-1-711-02-352
fax: 43-1-714-3582

Index

—A—

—B—

—C—

—D—

—J—

—L—

—N—

—Q—

—R—

—S—

YES, THE TOOLKIT WORKED FOR US AND WE WANT MORE INFORMATION

We found this toolkit a valuable resource for our ISO 9000 implementation strategies and documentation, and would like information about other programs that get the job done effectively and efficiently.

We would like to register to receive updates and information regarding the ISO 9000. Please send information regarding:

- ❐ ISO 9000 consulting and implementation program
- ❐ ISO 9000 in-house training
- ❐ ISO 9000 Quality System documentation workshop
- ❐ ISO 9000 technical writing services

Send information to:

____

Name

____

Company

____

Address

____

City, State, Zip

For more information, mail or fax this form to:
Quality Rite
Information Request
38 Hamlet St.
Newton, MA 02159
fax: 617-969-7273

YES, THE TOOLKIT WORKED FOR US AND WE WANT MORE INFORMATION

We found this toolkit a valuable resource for our QS 9000 implementation strategies and documentation, and would like information about other programs that get the job done effectively and efficiently.

We would like to register to receive updates and information regarding the QS 9000. Please send information regarding:

- ❐ QS 9000 consulting and implementation program
- ❐ QS 9000 in-house training
- ❐ QS 9000 Quality System documentation workshop
- ❐ QS 9000 technical writing services

Send information to:

____

Name

____

Company

____

Address

____

City, State, Zip

For more information, fax this form to:
Progressive Strategies & Systems, Inc.
fax: 810-781-4596
or
phone: 810-781-2235

LICENSE AGREEMENT AND LIMITED WARRANTY

READ THE FOLLOWING TERMS AND CONDITIONS CAREFULLY BEFORE OPENING THIS DISK PACKAGE. THIS LEGAL DOCUMENT IS AN AGREEMENT BETWEEN YOU AND PRENTICE-HALL, INC. (THE "COMPANY"). BY OPENING THIS SEALED DISK PACKAGE, YOU ARE AGREEING TO BE BOUND BY THESE TERMS AND CONDITIONS. IF YOU DO NOT AGREE WITH THESE TERMS AND CONDITIONS, DO NOT OPEN THE DISK PACKAGE. PROMPTLY RETURN THE UNOPENED DISK PACKAGE AND ALL ACCOMPANYING ITEMS TO THE PLACE YOU OBTAINED THEM FOR A FULL REFUND OF ANY SUMS YOU HAVE PAID.

1. **GRANT OF LICENSE:** In consideration of your payment of the license fee, which is part of the price you paid for this product, and your agreement to abide by the terms and conditions of this Agreement, the Company grants to you a nonexclusive right to use and display the copy of the enclosed software program (hereinafter the "SOFTWARE") on a single computer (i.e., with a single CPU) at a single location so long as you comply with the terms of this Agreement. The Company reserves all rights not expressly granted to you under this Agreement.

2. **OWNERSHIP OF SOFTWARE:** You own only the magnetic or physical media (the enclosed disks) on which the SOFTWARE is recorded or fixed, but the Company retains all the rights, title, and ownership to the SOFTWARE recorded on the original disk copy(ies) and all subsequent copies of the SOFTWARE, regardless of the form or media on which the original or other copies may exist. This license is not a sale of the original SOFTWARE or any copy to you.

3. **COPY RESTRICTIONS:** This SOFTWARE and the accompanying printed materials and user manual (the "Documentation") are the subject of copyright. You may not copy the Documentation or the SOFTWARE, except that you may make a single copy of the SOFTWARE for backup or archival purposes only. You may be held legally responsible for any copying or copyright infringement which is caused or encouraged by your failure to abide by the terms of this restriction.

4. **USE RESTRICTIONS:** You may not network the SOFTWARE or otherwise use it on more than one computer or computer terminal at the same time. You may physically transfer the SOFTWARE from one computer to another provided that the SOFTWARE is used on only one computer at a time. You may not distribute copies of the SOFTWARE or Documentation to others. You may not reverse engineer, disassemble, decompile, modify, adapt, translate, or create derivative works based on the SOFTWARE or the Documentation without the prior written consent of the Company.

5. **TRANSFER RESTRICTIONS:** The enclosed SOFTWARE is licensed only to you and may not be transferred to any one else without the prior written consent of the Company. Any unauthorized transfer of the SOFTWARE shall result in the immediate termination of this Agreement.

6. **TERMINATION:** This license is effective until terminated. This license will terminate automatically without notice from the Company and become null and void if you fail to comply with any provisions or limitations of this license. Upon termination, you shall destroy the Documentation and all copies of the SOFTWARE. All provisions of this Agreement as to warranties, limitation of liability, remedies or damages, and our ownership rights shall survive termination.

7. **MISCELLANEOUS:** This Agreement shall be construed in accordance with the laws of the United States of America and the State of New York and shall benefit the Company, its affiliates, and assignees.

8. **LIMITED WARRANTY AND DISCLAIMER OF WARRANTY:** The Company warrants that the SOFTWARE, when properly used in accordance with the Documentation, will operate in substantial conformity with the description of the SOFTWARE set forth in the Documentation. The Company does not warrant that the SOFTWARE will meet your requirements or that the operation of the SOFTWARE will be uninterrupted or error-free. The Company warrants that the media on which the SOFTWARE is delivered shall be free from defects in materials and workmanship under normal use for a period of thirty (30) days from the date of your purchase. Your only remedy and the Company's only obligation under these limited warranties is, at the Company's option, return of the warranted item for a refund of any amounts paid by you or replacement of the item. Any replacement of SOFTWARE or media under the warranties shall not extend the original warranty period. The limited warranty set forth above shall not apply to any SOFTWARE which the Company determines in good faith has been subject to misuse, neglect, improper installation, repair, alteration, or damage by you. EXCEPT FOR THE EXPRESSED WARRANTIES SET FORTH ABOVE, THE COMPANY DISCLAIMS ALL WARRANTIES, EXPRESS OR IMPLIED, INCLUDING WITHOUT LIMITATION, THE IMPLIED WARRANTIES OF MERCHANTABILITY AND FITNESS FOR A PARTICULAR PURPOSE. EXCEPT FOR THE EXPRESS WARRANTY SET FORTH ABOVE, THE COMPANY DOES NOT WARRANT, GUARANTEE, OR MAKE ANY REPRESENTATION REGARDING THE USE OR THE RESULTS OF THE USE OF THE SOFTWARE IN TERMS OF ITS CORRECTNESS, ACCURACY, RELIABILITY, CURRENTNESS, OR OTHERWISE.

IN NO EVENT, SHALL THE COMPANY OR ITS EMPLOYEES, AGENTS, SUPPLIERS, OR CONTRACTORS BE LIABLE FOR ANY INCIDENTAL, INDIRECT, SPECIAL, OR CONSEQUENTIAL DAMAGES ARISING OUT OF OR IN

Robin Short
Prentice Hall PTR
One Lake Street
Upper Saddle River, New Jersey 07458